Off The Grid
Preppers Power Survival Plan

Master DIY Solar & Wind Renewable Energy for Self-Sufficient Living and Cost Savings

By

Andrew Raines

Copyright © 2024 by Andrew Raines

All rights reserved.

No portion of this book may be reproduced in any form without written permission from the publisher or author, except as permitted by U.S. copyright law.

Contents

Introduction	1
1. Assess Your Energy Needs	3
2. Solar and Wind Power	13
3. Harness the Sun	33
4. Step-by-Step Solar DIY	59
5. Ride the Wind	73
6. Step-by-Step Wind Turbine DIY	91
7. Maintain Peak Performance	99
8. Alternative Energy Sources	113
9. Your Financial Roadmap	127
10. Navigate the Legal Landscape	145
Bonus Chapter 1: Generators	163
Bonus Chapter 2: The Complete Off-Grid Home	169
Conclusion	181
References	183

Introduction

Have you ever felt trapped by the endless cycle of utility bills, vulnerable to every hiccup in the power grid? Well, you're not alone. There's a world out there where the hum of generators is replaced by the whisper of the wind and the sun's warmth. In this world, independence isn't just a dream but a tangible, achievable reality. This isn't about turning your back on modern conveniences; it's about embracing a lifestyle that gives you control, freedom, and peace in an increasingly chaotic world.

With this book, I'm on a mission to share knowledge and empower you. We'll explore achieving energy independence through solar, wind, and other alternative energy strategies. We're talking practical, DIY solutions that will save you money and break the chains of dependency. From an in-depth analysis of power supply options to hands-on DIY guides for setting up your solar and wind power systems – I've got you covered. And yes, we'll venture into some of the lesser-known territories of off-grid energy solutions because who doesn't love a bit of adventure?

Whether you're a prepper, a DIY enthusiast, or someone looking to reduce your monthly bills and environmental footprint, you've picked up the right book. Perhaps it was the desire for a more sustainable lifestyle, or maybe the challenge of a DIY project that sparked your interest. Whatever your reasons, welcome aboard.

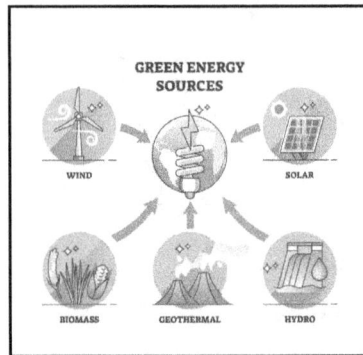

Green Energy

What sets this book apart is its foundation in professional engineering insights combined with practical, accessible guidance. This book is not just about the "how-to" but also the "why," focusing on well-known and innovative off-grid energy solutions.

Expect an engaging, empowering read that doesn't just talk at you but walks you through setting up your off-grid power system. Imagine completing projects that bring you closer to energy independence and give you a sense of accomplishment and freedom.

So, are you ready to take the first steps toward your off-grid aspirations? Let's embark on this journey together. Flip the page, and let's dive into the world of off-grid living, where every day is an opportunity to learn, grow, and take control of your energy future. Welcome to your *Off The Grid - Prepper's Power Survival Plan*.

1

Assess Your Energy Needs

Moving off-grid requires a clear understanding of your household's energy consumption. This isn't about rough estimates; it's about getting down to the brass tacks of how much energy you use, for what, and when. It's a critical first step that influences everything from the size of your solar array to your battery storage capacity.

Energy Audit: A DIY Guide

Embracing the off-grid lifestyle requires a keen awareness of energy consumption and an ongoing commitment to efficiency. Conducting an energy audit is a vital practice for off-grid enthusiasts as it provides a detailed understanding of where and how energy is used and, more importantly, how it can be conserved. Here, we delve deeper into the DIY auditing process, providing a step-by-step guide to uncovering and addressing energy inefficiencies.

Laying the Groundwork

Before you begin, it's crucial to have a clear plan and the right tools at your disposal. A successful DIY energy audit requires:

- *A Detailed Energy Consumption Record*: Gather data on your current energy usage. This could include past utility bills or your energy system's monitoring equipment readings.

- *An Energy Audit Toolkit*: Equip yourself with an energy meter, thermometer (or infrared camera), and a checklist of areas to inspect. Apps or devices that track energy usage in real time can also be invaluable.

- *A Blueprint of Your Living Space*: A floor plan can help you systematically approach each area and ensure every spot is audited.

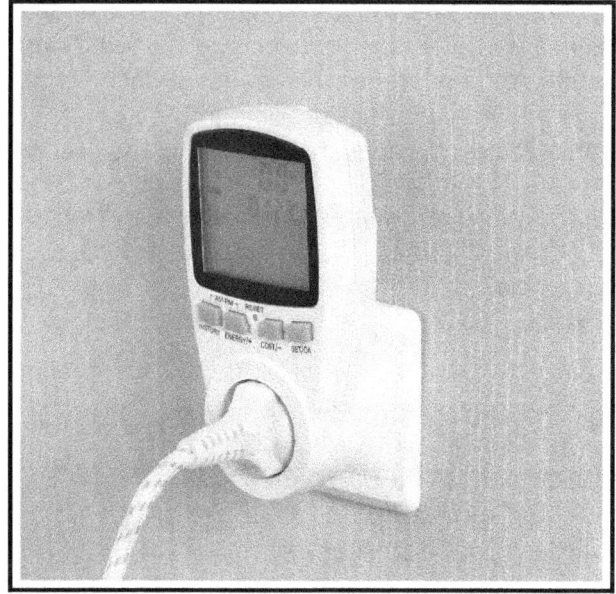
Energy Meter

Room Checklist

Lighting

- *Inventory All Light Sources*: Note the type and wattage of bulbs. Identify opportunities to switch to LED bulbs, which offer substantial energy savings over incandescent and halogen options.

- *Evaluate Usage Patterns*: Consider areas where motion sensors or timers could reduce energy waste, such as outdoor lighting or seldom-used spaces.

Appliances and Electronics

- *Measure Individual Appliance Energy Use*: Use a plug-in energy meter to understand how much power your appliances and devices consume. Your focus should be on those that draw the most energy. If you do not have a meter, then reference the appliance documentation for the energy consumption. This information is typically provided on the appliance label or in the user manual.

- *Identify Energy Vampires*: Look for devices that use power even when turned off, such as chargers, and consider using power strips to easily cut power when they're not in use.

Heating, Ventilation, and Air Conditioning (HVAC)

- *Check Filters and Ducts*: Clean or replace HVAC filters regularly and inspect ductwork for leaks, which can significantly impact system efficiency.

- *Assess Thermostat Settings*: Ensure your thermostat is set for efficiency and consider programmable options that adjust temperatures based on your schedule, like when you're out of the house for work.

Insulation and Windows

- *Inspect for Leaks*: Use your hand, a thermometer, or an infrared camera to find drafts around windows, doors, and other potential leak sites. Sealing these can dramatically improve heating and cooling efficiency.

- *Evaluate Insulation*: Check the attic, walls, and floors. Upgrading insulation where necessary can keep your home warmer in the winter and cooler in the summer, reducing the burden on your HVAC system.

Water Heating

- *Check Water Heater Settings*: Lowering the temperature setting on your water heater can save energy. So can insulating the water heater and the first six feet of hot and cold water pipes to reduce heat loss.

Standby Power

Standby consumption refers to the energy that appliances use when they are not actively in use or turned on. Most electrical appliances have a standby mode, where they may seem switched off but still consume power.

Globally, household standby consumption accounts for 2% of total electricity consumption and 1% of CO_2 emissions. Although this may not seem significant, the fact that this power is not being used or wasted makes it an issue worth addressing.

To avoid unnecessary power consumption by appliances on standby, turn them off at the power outlet or remove the plug.

Appliance Consumption Table

To help with your energy audit, here's a table of common household appliances. Just add up the power consumption for each appliance and the daily hours your total daily energy use.

Appliance	Category (Critical, Flexible, or Luxury)	No.	Power Rating (W)	Hours per Day	Daily Power Use (kWh)
Air Conditioner					
Clothing Iron					
Coffee Maker					
Cooktop and Oven					
Dishwasher					
Dryer					
Fan					
Freezer					
Games Console					
Hair Dryer					
Laptop					
Light Bulb					
Microwave					
Phone Charger					
Printer					
Refrigerator					
Slow Cooker					
Space Heater					
Toaster					
TV					
Vacuum Cleaner					
Washing Machine					
Other					

Analyze and Act

With your data in hand, prioritize improvements based on their potential energy savings and cost-effectiveness. Simple changes can offer immediate benefits, like switching to LED lighting or sealing leaks. At the same time, more extensive projects, like upgrading appliances or insulation, may require more investment but offer significant long-term savings.

Tracking Your Progress

After implementing changes, continue to monitor your energy consumption to gauge the effectiveness of your efforts. This ongoing process will help you fine-tune your energy usage and identify new opportunities for savings.

Leverage Subsidies and Incentives

After identifying potential improvements, explore subsidies and incentives that can help finance these upgrades. Resources like the Database of State Incentives for Renewables & Efficiency (DSIRE) provide comprehensive listings of available financial support options. This is explored in more detail in *Chapter 9: Your Financial Roadmap*.

Professional Energy Auditors

While this chapter focuses on DIY auditing, there are scenarios where hiring a professional energy auditor can add value. Professionals can offer detailed analyses, especially in complex systems or when eligibility for certain subsidies requires a certified audit. However, for many off-gridders, the primary pathway to energy efficiency lies in self-conducted audits, with professional services as a supplementary option.

Prioritize Needs

With a clear picture of your energy usage, the next step is to prioritize. Not all electrical devices are created equal. Some are essential (like the fridge), while others are luxuries (like a gaming console). Splitting your devices into 'needs' and 'wants' helps focus your off-grid system on the essentials, ensuring you have enough power for what's truly important.

Consider:

- *Critical*: These are your non negotiables. For most, that includes the fridge, essential lighting, and a water pump.

- *Meaningful but Flexible*: This category includes things like laundry machines or dishwashers, which are important. Still, you might use them less frequently or at off-peak times to save energy.

- *Luxuries*: Entertainment systems, decorative lights, or a hot tub fall here. Nice to have, but not critical.

Realistic Planning

Finally, when planning your off-grid system, it's crucial to be realistic. This means considering your daily energy needs and how those needs might change with the seasons or in emergency scenarios.

For example, energy consumption often increases in winter due to shorter days and the need for lighting and heating. Similarly, in summer, cooling systems can drive up energy use. Planning for these variations ensures you're not caught off guard.

Also, consider what might happen in an emergency. If you rely on a well for water, ensure your off-grid system can power the pump even in low sun or low wind conditions. This might mean having a larger battery bank or a backup generator.

Planning for these scenarios involves:

- *Seasonal Adjustments*: Consider how your energy needs change with the seasons and design your system accordingly. For example, this might mean more solar panels to account for shorter winter days.

- *Emergency Power*: Identify which systems must keep running in an emergency (like water pumps or medical devices) and ensure your system can support these, even in less-than-ideal conditions.

- *Flexibility and Redundancy*: Incorporating flexibility into your system, such as the ability to add more solar panels or batteries later, ensures it can evolve with your needs. Likewise, redundant systems (like a small backup generator) add an extra security layer.

Wrapping It Up...

Energy audits represent a cornerstone of the off-grid living philosophy, empowering individuals to take control of their energy usage and make informed decisions about improvements. By effectively combining these efforts with the strategic use of available subsidies and, where beneficial, the insights of professional auditors, off-grid residents can significantly enhance the efficiency and sustainability of their energy systems.

In pursuing a self-reliant lifestyle, the knowledge gained through DIY energy audits and the intelligent navigation of subsidy opportunities can lead to significant cost savings and environmental benefits, solidifying the foundation of your off-grid existence.

2

Solar and Wind Power

In the land of off-grid energy solutions, solar and wind power are the twin pillars of renewable energy. Heralded for their sustainability, accessibility, and efficiency, they are at the heart of renewable energy resources and play a pivotal role in pursuing off-grid independence.

Understand Solar Power

Imagine flipping a light switch in your home. That simple action connects you to one of humanity's most outstanding achievements: harnessing energy. Yet, the traditional path electricity takes to reach us is long, often dirty, and always dependent on a vast, vulnerable grid. Now, picture that same light switch being powered by the sun. The journey of that electricity is much shorter, cleaner, and under your control. Solar power isn't just about placing panels on your roof; it's about tapping directly into the universe's most abundant energy source.

At the heart of solar power is the photovoltaic (PV) cell, which converts sunlight into electricity without moving parts, noise, or emissions. It's like a green leaf, quietly and efficiently turning sunlight into usable energy. Here's how it works:

- Sunlight hits the PV cell, exciting electrons and creating an electric current.

- This current travels to an inverter, transforming it from DC (direct current) to AC (alternating current), the type of electricity used in homes.

- Electricity powers your home, and any excess can charge batteries for later use or, in some setups, be fed back into the grid for credits.

PV technology's beauty lies in its simplicity and efficiency. With solar panels now capable of converting more sunlight into electricity than ever before, the potential for solar energy has exploded.

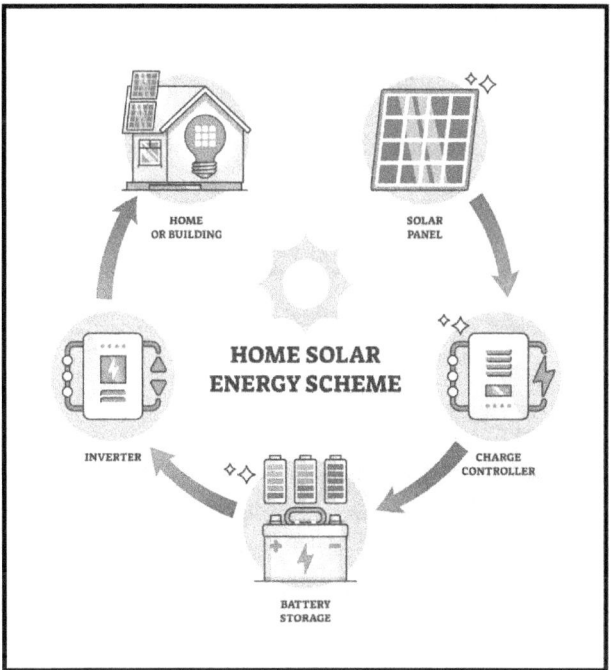

Off-Grid Solar Power

System Components

A solar power system is more than just panels. It's a symphony of components working in harmony, each with a crucial role:

- Solar panels capture sunlight. Think of them as the frontline soldiers in the battle for energy independence.

- Inverters convert the DC electricity from solar panels into AC electricity. They're the translators, making the energy speak a language your home understands.

- Batteries store excess energy. Like squirrels with their nuts, batteries ensure you have power even when the sun isn't shining.

- Charge controllers protect your batteries by regulating the flow of electricity, ensuring they're neither overcharged nor drained too low.

Each component is vital, and understanding how they fit together gives you the power to tailor your system to your needs, ensuring reliability and efficiency.

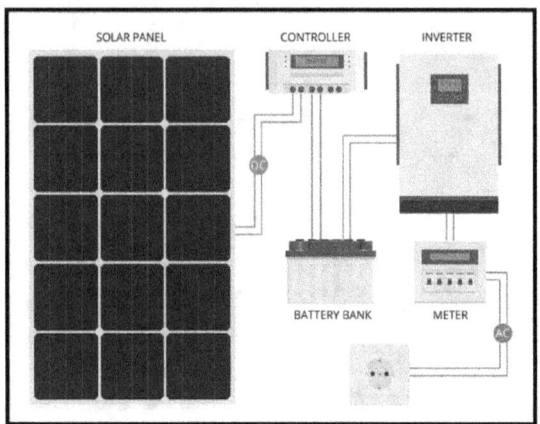

Off-Grid Solar System Components

Benefits and Limitations

Solar energy is like a coin with two sides. On one hand, the benefits are clear:

- *Renewable and Abundant*: The sun sends enough energy in an hour to Earth to power our global needs for a year.

- *Low Maintenance*: Once installed, solar systems require little upkeep. They're the "set it and forget it" of the energy world.

- *Sustainability*: Solar power reduces reliance on fossil fuels, cuts carbon emissions, and helps mitigate climate change.

On the flip side, solar energy has its challenges:

- *Initial Costs*: The upfront investment in solar can be steep, though falling component prices and incentives make it more accessible.

- *Sunlight Dependency*: Solar power relies on the sun, making it less predictable than traditional power sources. This variability necessitates batteries or alternative energy sources to fill the gaps.

Understanding these aspects helps set realistic expectations and plan more effectively for a resilient off-grid system.

Advancements in Solar Technology

The solar industry is not standing still. It's sprinting. Recent innovations are making solar power more efficient, affordable, and adaptable:

- *Higher Efficiency Panels*: New materials and designs are pushing the boundaries of how much sunlight panels can convert into electricity.

- *Flexible Solar Panels*: Imagine rolling out a solar panel like a yoga mat on various surfaces, from RVs' roofs to boats' sails. That's the promise of flexible panels.

- *Solar Tiles*: These blend into your roof, offering a more aesthetically pleasing option that turns your entire roof into an energy generator.

- *Portable Solar*: From backpacks with built-in panels to foldable arrays for camping, solar power is going places it's never been before.

These advancements are about more than just generating power. They're about integrating that power into our lives more seamlessly and efficiently. Solar technology is becoming less about what it is and more about what it can do for us.

As we peel back the layers of solar power, from its basic principles to the cutting-edge technologies reshaping its future, we're not just discussing an alternative to traditional energy sources. We're exploring a revolution in how we think about and use energy. Solar power, with its blend of benefits and challenges, simplicity and complexity, stands at the forefront of this revolution, offering a path to a cleaner, more sustainable, and more independent future.

With each advancement and every system installed, we're not just turning sunlight into electricity. We're turning rooftops into power plants, homeowners into energy producers, and dreamers into doers. The bright solar path leads us toward a future where energy is consumed, generated, shared, and celebrated.

Harness the Breeze

Wind energy taps into the natural flow of air across our planet, transforming kinetic energy into a form we can use – electricity. It's a process that aligns closely with the rhythms of nature, offering a renewable source of power that complements the sun's bounty. Understanding wind power begins with recognizing its core components and the fundamental principles that allow a gust of wind to light up a home.

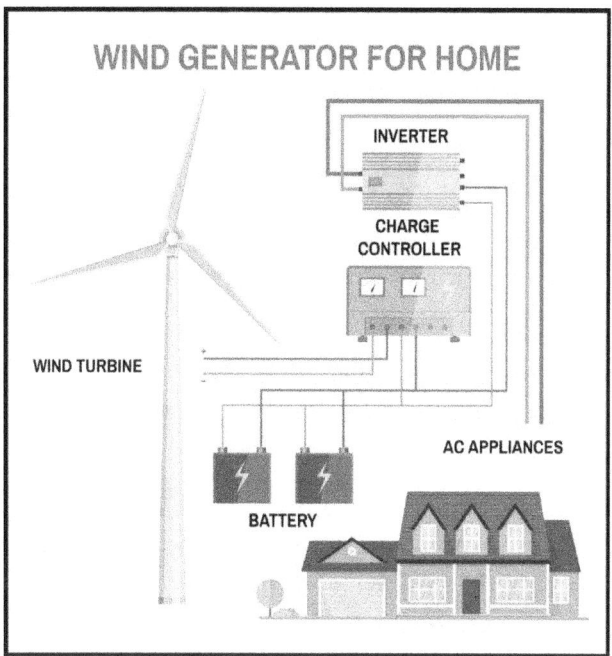

Off-Grid Wind Turbine Power

System Components

Wind turbines are energy workhorses that stand tall, capturing the breeze with their blades, which act much like an airplane's wings. Air pressure differences across the blades cause them to spin, driving a generator that produces electricity. This setup comprises several key elements:

- Blades to catch the wind.

- Rotor to turn blade motion into rotational energy.

- Generator to convert rotational energy into electricity.

- Tower to elevate the rotor and blades to higher, stronger winds.

Each component plays a crucial role in efficiently converting wind to electricity, showcasing the elegance of wind power technology.

Benefits and Limitations

Wind energy offers several compelling advantages, particularly when integrated into off-grid living setups. Its main strengths include:

- *Renewable and Abundant*: Wind is a limitless resource generated by the heating and cooling of the atmosphere, the rotation of the Earth, and the planet's geography.

- *Reduced Carbon Footprint*: Like solar, wind energy produces no emissions during operation, contributing to a cleaner, more sustainable energy future.

- *Complementarity With Solar Power*: Wind speeds tend to be higher when solar resources are low, such as during cloudy days or at night, making wind and solar a potent combination for off-grid systems.

However, wind energy also faces challenges that must be navigated:

- *Variability*: Wind is not constant, and its unpredictability can lead to periods of low energy production.

- *Noise*: Turbines produce sound as their blades cut through the air, which can be a consideration if living in a residential area.

- *Impact on Wildlife*: Birds and bats can be at risk, though careful site selection and design can minimize these impacts.

Incorporating wind energy into an off-grid power system requires careful planning and consideration of both the benefits and limitations of wind power. You can make informed decisions that enhance your off-grid setup by understanding how wind turbines work, assessing your property for wind potential, and weighing the pros and cons of wind energy. With the right approach, wind energy can provide a reliable, renewable source of power that complements other off-grid energy solutions, bringing you one step closer to energy independence.

Site Evaluation for Solar and Wind

Finding the perfect spot on your land for solar panels or a wind turbine goes beyond just picking a place with a nice view. It's about understanding where the elements of nature work best in your favor. This part of your off-grid adventure requires observation, research, and, sometimes, a little bit of technology to get the most accurate information.

Solar Insolation and Sun Path

At first glance, solar insolation might sound like a complex term. However, it's simply a measure of how much solar power you will get in a specific area over time. Think of it as measuring how generously the sun bathes your site in its energy. Knowing this helps you pinpoint the optimal location for your solar panels to soak up the sun.

- *Measure Solar Insolation*: Various online tools and maps can give you a ballpark figure of the solar insolation in your area. These resources compile years of weather and satellite data to provide average monthly and annual solar energy figures tailored to your location.

- *Understand the Sun's Path*: Observing how the sun travels across your property throughout the day and seasons is invaluable. The goal is to find spots that enjoy uninterrupted sunshine for the longest periods, especially during peak solar hours (typically between 9 a.m. and 3 p.m.). Trees, buildings, or even the slope of the land can affect this, so keep an eye out for potential shadows.

Wind Speed and Turbulence

Wind energy's potential hinges on two critical factors: speed and turbulence. Wind speed determines how much power your turbine can generate. In contrast, turbulence, caused by obstacles disrupting airflow, can affect turbine efficiency and lifespan.

Deciding where to place a wind turbine on your property is a crucial step that can significantly impact its performance. Wind patterns can vary dramatically even within the same property as they're influenced by terrain, structures, and vegetation. Assessing a site for wind energy potential involves a few critical steps:

- *Assessing Wind Speed*: Observing local wind patterns might include consulting wind maps or data from nearby weather stations to understand prevailing wind directions and speeds. Monitoring wind speed over a year using an anemometer clearly shows your site's wind energy potential.

- *Understanding Turbulence*: Identifying potential sites where wind flow is unobstructed by trees, buildings, or topographical features is crucial. Open, flat spaces usually offer less turbulence, making them better suited for wind turbines.

- *Considering Elevation*: Wind speeds increase with height, making taller towers more effective at capturing energy.

Combining Solar and Wind

Integrating solar and wind power can create a more stable and reliable off-grid energy system, covering the bases when one source is not at its peak. The trick is to understand how they can complement each other. Here are two things you can do:

- *System Integration*: The integration requires a setup that can seamlessly switch between or combine the two energy sources. This often involves dual charge controllers, one for each system, and an inverter that can handle input from both.

- *Balancing the Load*: Analyzing your energy consumption patterns helps balance how much energy you draw from each source. For instance, if your area has windy nights and sunny days, your system can prioritize wind energy for nighttime use and solar power for the day.

Environmental Considerations

While exploring the potential for solar and wind energy on your property, it's crucial to tread lightly on the land and its inhabitants. There are two things you should look out for:

- *Land Usage*: Solar panels and wind turbines require space. When choosing locations for these installations, consider how they'll impact the land use. For solar panels, rooftops or already cleared land might be ideal to minimize disruption. Wind turbines should be placed where they have minimal impact on the landscape and soil.

- *Wildlife Impact*: Both solar farms and wind turbines can affect local wildlife. For solar installations, consider elevated panels to allow ground animals free movement. For wind turbines, careful placement can minimize risks to birds and bats. Engaging with local environmental groups can provide insights into the best practices for reducing these impacts.

In the quest for off-grid energy independence, aligning with the forces of nature is vital. By thoughtfully evaluating your site for solar and wind potential, considering both the generous gifts and the environmental challenges, you set the stage for a harmonious and sustainable off-grid lifestyle. This process, rooted in respect for the natural world and a commitment to stewardship, ensures that your off-grid journey enriches your life and the land and ecosystems that support it.

Do You Need Battery Storage?

Battery storage plays a crucial role in off-grid solar or wind turbine systems due to the inherent nature of renewable energy sources: They are intermittent and don't always produce power at times of demand. Here's why battery storage is essential for anyone looking to be off-grid.

Energy Availability

- *Intermittency*: Solar and wind energy production is subject to weather conditions and time of day. Solar panels only generate electricity during daylight hours, and their output varies with the weather and seasons. Similarly, wind turbines depend on wind speed, which can be unpredictable.

- *Constant Demand*: Energy consumption, on the other hand, occurs around the clock, with peaks typically in the morning and evening when solar and wind generation might be low or nonexistent.

- *Storage Solution*: Batteries store excess energy produced during peak generation times, making it available for use during periods of low production or high demand, ensuring a continuous energy supply.

Energy Independence

- *Off-Grid Living*: For homes not connected to the utility grid, battery storage is essential for providing a reliable power source that can meet all energy needs without an external supply.

- *Self-Sufficiency*: Battery storage allows off-grid systems to operate independently, reducing reliance on traditional energy sources and enhancing energy security.

Optimizing Energy Use

- *Load Shifting*: Battery storage enables shifting energy consumption from times of low production to times of high production. By storing energy when it is abundantly produced and using it during shortages, households can optimize their energy use according to production patterns.

- *Increased Efficiency*: Properly sized battery storage can reduce waste by capturing surplus energy that would otherwise be lost due to overproduction during periods of low demand.

System Stability and Reliability

- *Voltage and Frequency Regulation*: Battery systems can help stabilize your off-grid system by regulating voltage and frequency, ensuring that the power supplied to your home is within safe and usable limits.

- *Emergency Backup*: In case of equipment failure or extremely low production due to adverse weather conditions, batteries serve as an emergency power reserve, maintaining critical loads until production resumes.

Financial Efficiency

- *Reduced Costs*: While the initial setup cost for battery storage can be high, the long-term savings on electricity bills and the avoidance of grid connection or diesel generator costs can make it financially efficient over time.

- *Investment in Sustainability*: Investing in battery storage is also an investment in sustainability, reducing your carbon footprint by maximizing the use of renewable resources.

In summary, battery storage is indispensable for off-grid solar or wind systems as it provides a buffer that reconciles the mismatch between renewable energy production and household energy consumption patterns. It ensures a reliable, continuous power supply, fosters energy independence, and supports a sustainable lifestyle.

Battery Types, Considerations, and Future

Understanding the different types of batteries, how to size your battery bank, and the importance of maintenance and safety can enhance the resilience of your off-grid living setup. Let's also peek into what the future holds for battery technologies.

Types of Batteries

In off-grid systems, batteries are the silent guardians of your energy supply, ready to step in when the elements rest. The most common types you'll encounter are:

- *Lead-Acid*: These are the veterans of battery technology, valued for their reliability and cost-effectiveness. They come in two main varieties: flooded (which requires regular maintenance) and sealed (which are maintenance-free but more expensive). However, they're heavy, have a shorter lifespan than other types, and require careful handling due to their toxic lead and acid electrolyte content.

- *Lithium-Ion*: These are the new kids on the block. They are lighter, have a higher energy density, offer longer lifespans, and are becoming increasingly popular for off-grid systems due to their efficiency and declining cost. The main drawbacks are their upfront cost and more complex management systems to ensure safety and longevity.

- *Saltwater*: A newer entry into the battery market, these batteries use a salt solution as their electrolyte. They're nontoxic, making them an environmentally friendly option. While they offer lower energy density and are relatively new (meaning less historical data on longevity and performance), their safety and sustainability make them an intriguing option for off-grid systems.

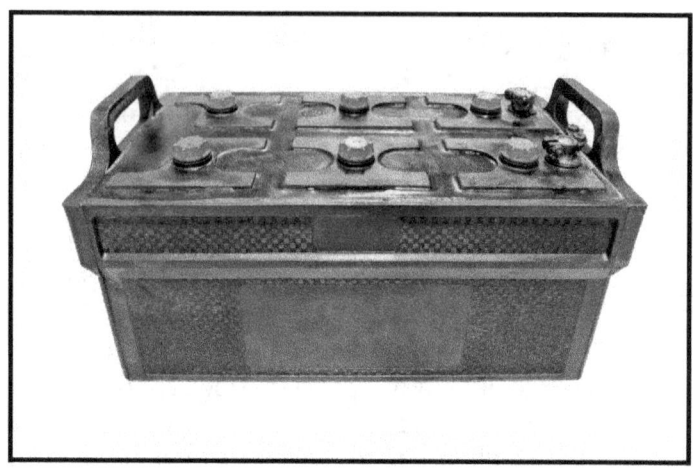

Lead acid Battery

Size Your Battery Bank

Determining the right size for your battery bank is a balancing act. Too small, and you risk running out of power; too large, and you're overspending on capacity you don't need. The key is to understand your energy usage patterns and size your battery bank to meet your needs, considering:

- *Daily Energy Consumption*: This is the daily energy usage, in kilowatt-hours, that you calculated during your DIY energy audit in *Chapter 1*.

- *Rated Power Output*: This is the total electricity load your battery can supply simultaneously in kilowatt-hours.

- *Storage Capacity*: This is the total capacity of the battery in kilowatt-hours. For example, a battery with a storage capacity of 10 kWh and a rated power output of 3 kW is not fit for purpose if your simultaneous demand is higher than 3 kW.

- *Daylight Hours Consumption*: On average, households use about a third of daily power consumption during daylight hours. A storage capacity of approximately two-thirds of the daily consumption is required to meet the nighttime supply demand.

- *Depth of Discharge (DoD)*: Batteries last longer when not completely drained. Most batteries have a recommended DoD, usually around 50% for lead-acid and up to 80% for lithium-ion, which should be factored into your calculations.

- *Efficiency*: Batteries lose some energy through heat during charging and discharging. Accounting for this inefficiency is crucial when sizing your battery bank.

Lithium-ion Battery

Energy Contingency

To power your devices at night, a battery needs the capacity to supply approximately two-thirds of your daily consumption. For example, if a household has a daily usage of 15 kWh, then a minimum of 10 kWh (two-thirds of 15 kWh) of battery storage is required for nighttime consumption. The battery will also need a rated power output to meet simultaneous demand.

If your budget allows, increasing battery storage capacity can be beneficial; however, even though costs have decreased significantly over the past few years, batteries still require a considerable investment. Learn more about costs in *Chapter 9: Your Financial Roadmap*.

Future Battery Technologies

The future of battery technology is bright, with innovations promising to make off-grid living even more viable and sustainable. Here are a few new batteries pushing batter technology into the future:

- Graphene batteries are on the horizon, offering the potential for faster charging times, higher capacities, and longer lifespans.

- Solid-state batteries eliminate the liquid electrolyte found in lithium-ion batteries, aiming for even higher energy densities and safety levels.

- Research into organic and biodegradable batteries presents a future where battery disposal no longer poses environmental hazards.

As these technologies mature, they promise to reduce costs, reduce environmental impact, and improve the performance of off-grid energy systems.

Grid-Tied vs. Off-Grid Systems

To grid or not to grid! Most of us are looking at the benefits of an off-grid system and the independence it provides; however, connecting to the grid does offer some advantages. Here is a side-by-side comparison of tying into the grid or remaining completely independent:

- *Grid-Tied Systems*: These systems allow you to sell excess energy back to the grid while drawing from it when your energy production is low. The beauty of this setup is the financial savings and potential earnings from the energy you feed back. You can also avoid the highest kilowatt-hour prices on a time-of-use tariff or when the price of electricity changes depending on the time of day. However, grid-tied systems can be complex, requiring synchronization with the utility grid and adherence to regulations.

- *Off-Grid Systems*: An off-grid system is the way to go for those seeking true independence from utility companies. The challenge here lies in ensuring your system can meet all your energy needs without the grid as a backup. This often means a more significant upfront investment in turbines, solar panels, and batteries but results in full autonomy over your power supply.

Choosing between grid-tied and off-grid configurations hinges on your goals, budget, and the level of energy independence you're aiming for.

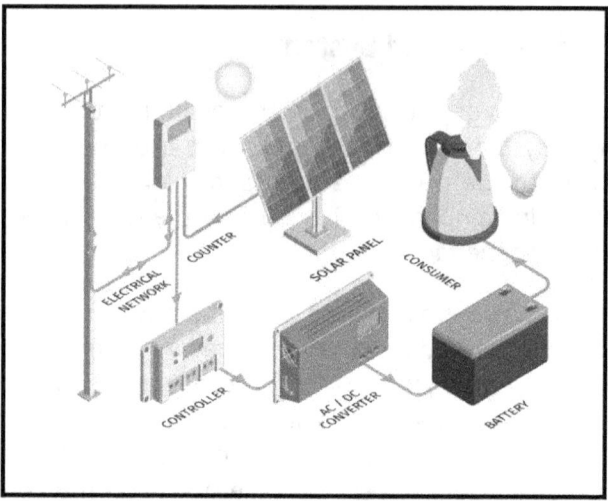

Grid-Tied Solar Power

Wrapping It Up...

In this chapter, we've discussed the advancements and efficiency of solar and wind energy, emphasizing the importance of a balanced approach to off-grid energy systems, including evaluating potential sites and integrating battery storage solutions. Understanding these renewable sources can help individuals harness them effectively, paving the way toward a sustainable and energy-independent future.

As we turn the page, we'll explore the practical steps and considerations for installing your solar power system, ensuring you're equipped to bring the power of the sun right to your doorstep.

3

Harness the Sun

In this chapter, we embark on a detailed exploration of solar power, looking into the ecosystem of components that collectively form an efficient solar power system. Each element is critical in effectively capturing, storing, converting, and utilizing solar energy. Through a comprehensive examination of these individual elements, we aim to provide you with a deeper understanding of how solar power systems function and how to optimize their performance.

Panel Types and Efficiency

When it comes to solar panels, one size doesn't fit all. The market offers a variety of types, each with its unique features, benefits, and best-use scenarios. Let's break them down:

- *Monocrystalline Panels*: These solar panels are made from a single, pure silicon crystal and are considered to be the most efficient. Due to their high-grade silicon, they can convert more sunlight into electricity and have a sleek black design. They are perfect for small spaces because of their high power output and typically have efficiencies ranging from 18% to 22%. However, these panels do come with a higher price tag.

- *Polycrystalline Panels*: Polycrystalline panels are commonly used in the solar industry and are made by melting silicon fragments together. They have a blue hue and somewhat speckled appearance and are slightly less efficient than their monocrystalline counterparts. However, they offer a balance of performance and cost, which makes them a popular choice for many off-grid solar applications. Usually, their efficiency ranges from 15% to 17%.

- *Thin-Film Panels*: Thin-film panels are a unique type of solar panel that differs from monocrystalline and polycrystalline panels. They are made by layering PV material on a substrate, which makes them light in weight, flexible, and suitable for applications where traditional panels might not work. However, they have a lower efficiency and require more space to produce the same amount of electricity as other types of solar panels. Generally, they are the least efficient option for residential use, with efficiencies ranging from 10% to 13%.

Different types of solar panels have their own unique advantages depending on various factors like climate, space availability, and energy requirements. For instance, using monocrystalline panels can be the most efficient option if you live in less sunlight. On the other hand, if you have enough space and are looking for a cost-effective solution, polycrystalline or thin-film panels could be the way to go.

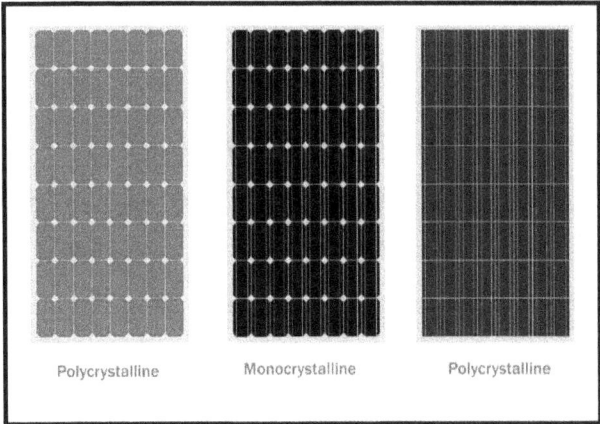

Solar Panel Types

The Future of Solar: Cutting-Edge Technologies and Materials

Solar energy technology has seen tremendous advancements in recent years, pushing the limits of efficiency, affordability, and application flexibility. New materials and innovations are set to revolutionize the solar panel market, offering improved energy capture and integration capabilities. Below are some of the most exciting developments in the solar industry:

Perovskite Solar Cells: The Next-Generation Efficiency Boost

Perovskite solar cells are emerging as one of the most promising breakthroughs in solar energy. Unlike traditional silicon panels, perovskites use a unique crystalline structure that is highly efficient at capturing sunlight. Current lab-tested efficiencies have reached over 30%, surpassing traditional silicon panels.

Advantages:

- Higher efficiency potential compared to silicon panels
- Lightweight and flexible design
- Can be printed on various surfaces, enabling integration into windows and walls
- Lower production costs compared to silicon panels

Challenges:

- Stability and durability issues—current versions degrade faster than silicon panels
- Ongoing research is needed to improve lifespan and commercial scalability

Many experts believe that perovskite-silicon tandem solar cells—which combine both materials—will be the first commercially viable perovskite application, offering efficiency levels above 30%.

Bifacial Solar Panels: Capturing Sunlight from Both Sides

Bifacial solar panels have been gaining popularity because they can capture sunlight from both the front and back sides, utilizing reflected light to increase energy production by 10–20%.

Best Applications:

- Ground-mounted systems where reflective surfaces (like white roofs or desert sands) boost rear-side capture
- Areas with high sunlight exposure to maximize efficiency

Bifacial panels are becoming increasingly cost-effective as manufacturers improve production methods. They are especially attractive for off-grid users who need to maximize efficiency in limited spaces.

Tandem Solar Cells: Stacking Efficiency Layers

Tandem solar cells stack multiple light-absorbing layers to capture a broader spectrum of sunlight. By combining different materials—such as perovskite and silicon or gallium arsenide—they can push efficiencies beyond 40% in lab settings.

Why They Matter:

- Higher power output from the same surface area
- Potential to lower costs per watt of electricity generated
- Improved performance under varying light conditions

While still in development, tandem solar cells could reshape the solar industry as production scales up.

Transparent Solar Panels: Turning Windows into Energy Sources

One of the most futuristic developments in solar energy is transparent solar technology. These panels can be applied to windows, car windshields, and even smartphone screens to generate electricity while maintaining visibility.

How They Work:

- Transparent solar panels use organic photovoltaic (OPV) technology or ultra-thin perovskite layers to absorb non-visible light while allowing visible light to pass through.

- They can turn skyscrapers, residential windows, and greenhouse glass into power-generating surfaces.

Companies like Ubiquitous Energy are working to commercialize this technology, with expected rollouts in the next 5–10 years.

Solar Paint and Coatings: Turning Any Surface into a Power Generator

Solar paint is an emerging technology that could eliminate the need for traditional panels altogether. Scientists are developing solar coatings that can be sprayed onto surfaces to absorb sunlight and convert it into electricity.

Potential Applications:

- Cars and boats with solar-painted exteriors
- Building walls that generate electricity
- Remote locations where installing panels is impractical

While still in the experimental phase, solar paint could be a game-changer for off-grid energy solutions in the near future.

Quantum Dot Solar Cells: Ultra-Lightweight and Highly Efficient

Quantum dot solar cells use nanotechnology to fine-tune the absorption of sunlight at the molecular level, allowing for unprecedented efficiency improvements.

Why They're Exciting:

- They can achieve efficiencies beyond 40% in lab conditions.

- Extremely lightweight and flexible, allowing for wearable solar technology.

- Potential to significantly lower manufacturing costs once commercialized.

Though still in development, quantum dots represent a potential leap forward in solar technology.

What's Coming Soon?

Several of these next-gen solar technologies are expected to become commercially available in the next 3–7 years. Perovskite solar cells and bifacial panels are already being introduced into the mainstream market, while transparent panels and solar coatings are on the horizon.

What This Means for Off-Grid Solar Users

Increased Efficiency: New materials will allow higher power output from smaller panels, making solar energy even more viable in limited spaces.

Lower Costs: As technology advances, production costs will drop, making solar energy more affordable than ever.

More Integration Options: With solar coatings and transparent panels, solar power won't be limited to rooftops—everything from windows to car exteriors could generate energy.

For off-grid living, these cutting-edge technologies mean that solar energy will soon be more efficient, more accessible, and more versatile than ever before.

With the continued advancements in solar panel technology, the future of off-grid energy is looking brighter and more sustainable. Whether you choose traditional monocrystalline panels or opt for emerging innovations like perovskite or bifacial panels, the options for self-sufficient power generation are expanding rapidly.

Household Power Requirements

According to the most recent data from the U.S. Energy Information Administration (EIA), a U.S. household's average daily electricity consumption is 29 kWh. This figure is for households connected to the grid receiving electricity from utility companies. Off-grid households tend to consume less electricity as off-gridders are more conscious and efficient in their power usage.

Knowing how much power you need is crucial before you go panel shopping. Here's a quick way to get a rough estimate:

- Calculate your daily energy use in kilowatt-hours. Use the results from your DIY energy audit in *Chapter 1*.

- Consider the average peak sunlight hours your location receives. You can find information on peak sunlight hours for states and major cities from the National Renewable Energy Laboratory (NREL). The average peak hours for the U.S. is 4.5 hours.

- Use these figures to calculate how many watts your solar panels will need to produce to meet your energy demands.

Remember, it's not just about covering your current usage. Think about future needs, too. Are you planning on adding an extra room or buying an electric vehicle? Factor these into your calculations.

Your solar system must produce the kilowatt-hours corresponding to your daily energy usage to cover most electricity needs.

For example, suppose we use a household power consumption of 15 kWh, approximately half of the average grid-tied US household consumption of 29 kWh, and 4.5 hours of peak sunlight daily. In that case, we get the following solar power requirement:

Solar Power (kW) = Daily Energy Use (kWh) / Average Daily Sunlight (hrs)

=   15 kWh / 4.5 hrs

=   3.3 kW

Therefore, to meet your household's daily power requirements of 15 kWh, your solar panels must produce 3.3 kW of energy for 4.5 hours daily. This assumes you can store the excess energy in batteries to supply your household during the nighttime or on days with low sunlight.

Additionally, when choosing the size of your solar power system, you need to consider the system's inefficiencies and battery storage. Solar panels may not always perform at their highest output due to shading, dust, and temperature changes. Experts recommend adding a 25 percent buffer to your target daily average to account for these issues and ensure you can produce all the clean energy you need. This would increase the 3.3 kW power requirement for our solar panels in our example to 4.13 kW.

How Many Solar Panels Do You Need?

Once you have determined your solar power requirements, you can calculate the number of required solar panels. To do this, you need to measure the physical size of your available space. Suppose your roof area is limited or partially shaded. In that case, you should use fewer, smaller, high-efficiency panels that can generate the most power over time. On the other hand, if you have sufficient space, you can compromise some efficiency by purchasing larger panels at a lower cost per panel to reach your desired energy output.

Durability and Warranties

Solar panels can endure extreme weather and come with warranties, making them a wise investment. Look for:

- *Performance Warranties*: These guarantee the panels will produce a certain percentage of their rated power after several years.

- *Product Warranties*: These cover defects in manufacturing or materials.

A solid warranty can be a tie-breaker if you're stuck between two options. It's not just about peace of mind; it's about protecting your investment.

Mounting Solutions for Maximum Efficiency

Selecting the ideal spot and method for placing your solar panels isn't just a matter of aesthetics or convenience; it's about capturing every possible ray of sunlight and converting it into usable energy. Whether you're working with a sprawling backyard, a cozy rooftop, or a unique setup requiring creative thinking, there's a mounting solution to fit your needs.

Mounting Options

The choice of mounting plays a pivotal role in the performance of your solar setup. Here's a look at the primary options:

- *Ground Mounts*: These systems plant your solar panels firmly on terra firma. Ground mounts are ideal for those with ample yard space as they are easily accessible for maintenance and are adjustable so you can easily angle them toward the sun. They can, however, require significant land clearing and preparation work.

- *Roof Mounts*: For many, the roof is prime real estate for solar panels. It's an unobtrusive option that takes advantage of existing structures, saving ground-level space for other uses. The key is ensuring your roof can bear the weight and is in good repair before installation.

- *Pole Mounts*: Elevating your solar panels on poles can lift them above shading obstacles and position them perfectly toward the sun. This option can be more expensive due to the need for sturdy poles and potentially complex installation, but it is excellent for maximizing exposure.

Choosing between these options involves weighing factors like space availability, budget, and environmental conditions. Each has its advantages, and sometimes, a combination of mounting types is the best approach to meet all your energy needs.

Ground Mounted Solar Panels

Roof Mounted Solar Panels

Orientation and Angle

Getting the most out of your solar panels means aligning them with the sun's path as closely as possible. Here's what to consider:

- *Orientation*: In the northern hemisphere, solar panels should face true south for maximum sunlight exposure. For those in the southern hemisphere, it's the opposite, with panels best oriented toward true north.

- *Angle*: The tilt of your panels should mirror your latitude for optimal year-round sunlight capture. Adjustments can be made to favor summer or winter sun if your energy needs fluctuate seasonally.

Fine-tuning the orientation and angle of your solar panels can dramatically improve their efficiency and, by extension, the overall output of your solar power system.

Pole Mounted Solar Panels

Materials and Durability

The materials used in your mounting system are the unsung heroes, quietly ensuring that your solar panels remain secure and effective through all sorts of weather. Here's what to look for:

- *Corrosion Resistance*: Materials like aluminum and stainless steel stand up well against the elements, resisting rust and degradation over time.

- *Strength and Durability*: Your mounting system should be robust enough to support the weight of your panels and withstand environmental stressors like high winds and heavy snow loads.

Investing in quality materials for your mounting system is investing in the longevity and reliability of your solar power setup.

Charge Controllers

In the dynamic world of off-grid solar power, where the sun's energy is converted into electricity to light up homes and power gadgets, an unsung hero, the charge controller, is working quietly behind the scenes. This device might not grab headlines, but its role is pivotal in safeguarding your solar power system, ensuring the batteries that store this precious energy remain healthy and efficient throughout their lifespan.

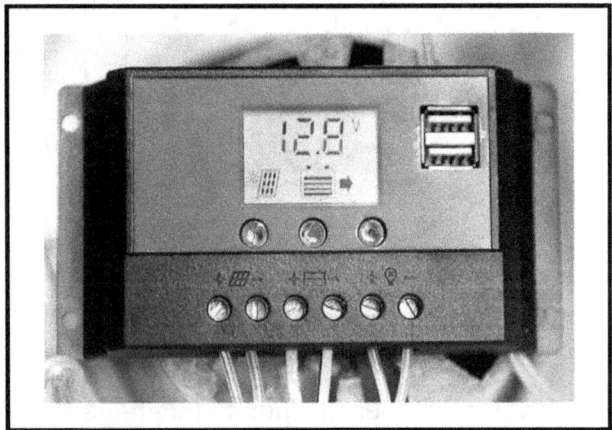

Charge Controller

The Role of Charge Controllers

Imagine your solar system as a bustling city. In this city, electricity flows like traffic, and the charge controller is the traffic cop, directing this flow to where it's needed and preventing any potential gridlock, which means overcharging your batteries. Batteries are the backbone of any off-grid solar system, storing energy when the sun isn't shining. However, they have their limits. Push too much energy into them too quickly, or try to cram in more than they can hold, and you risk shortening their life or, worse, causing damage. The charge controller modulates the flow of electricity, ensuring batteries are charged safely and efficiently, thus extending their life and protecting your investment.

Pulse Width Modulation (PWM) and Maximum Power Point Tracking (MPPT)

Regarding charge controllers, you'll encounter two main types: pulse width modulation (PWM) and maximum power point tracking (MPPT). Choosing between them is akin to selecting between a reliable sedan and a high-performance sports car; each has advantages, depending on your needs and budget.

- *PWM Controllers*: These are the workhorses of the charge controller world. Simple yet effective, PWM controllers slow the flow of electricity from the panels to the batteries as they get closer to full charge, reducing the risk of overcharging. They're cost-effective and work well with systems where the solar panel voltage closely matches the battery voltage. Ideal for smaller setups or those on a tight budget, PWM controllers offer a balance between performance and cost.

- *MPPT Controllers*: These high performers are equipped with the technology to maximize the energy harvested from your solar panels. MPPT controllers adjust their input to capture the maximum power from the panels, converting excess voltage into amperage. This process can significantly increase the efficiency of your solar system, especially in colder temperatures or when the panel voltage is much higher than the battery voltage. Though more expensive, MPPT controllers are perfect for larger systems or those looking to squeeze every bit of power out of their solar panels.

Choose the Right Controller

Selecting the perfect charge controller for your solar setup isn't a decision to make lightly. It involves considering not just the size of your system but also the specific characteristics of your panels and batteries and your budget. Here are a few factors to weigh:

- *System Size*: Larger systems with high-voltage panels are better paired with MPPT controllers to exploit their efficiency-boosting capabilities. Smaller, more straightforward setups might see less benefit, making a PWM controller a cost-effective choice.

- *Battery Type*: Different batteries have different charging profiles. Some, like lithium-ion batteries, require precise management to maximize their lifespan, something MPPT controllers excel at.

- *Budget Constraints*: While MPPT controllers offer higher efficiency, they also have a higher price tag. If your system is on the smaller side or budget is a concern, a PWM controller might be all you need.

Set and Adjust

Getting your charge controller settings right is like fine-tuning an instrument. It's about hitting those perfect notes that balance battery health with system efficiency. Most controllers come with pre-set modes tailored to common battery types. However, diving into manual adjustments can unlock even better performance. Here's how:

- *Understand Your Batteries*: Start by knowing your batteries' charging specifications. This includes the bulk, absorption, and float voltages the manufacturer recommends.

- *Adjust According to Conditions*: Factors like temperature can affect battery charging. Some controllers come with temperature sensors, allowing for automatic adjustments. If yours doesn't, you might need to tweak settings manually as conditions change.

- *Monitoring and Tweaking*: Keep an eye on your system's performance. Modern charge controllers offer monitoring through digital displays or even smartphone apps, providing insights into how well your batteries are charging and when adjustments might be needed.

In essence, charge controllers are more than just gatekeepers for your batteries. They're crucial to optimizing the performance and longevity of your off-grid solar system. By understanding their role, choosing between PWM and MPPT based on your specific needs, and fine-tuning settings for optimal charging, you ensure your solar setup isn't just efficient but also built to last.

Battery Storage for Solar Power

Storing today's sun for an overcast tomorrow is where battery storage shines in a solar energy setup. Batteries act as a bank, holding on to the energy generated by your solar panels until it's time to light up your home, power your devices, or keep your refrigerator running. Batteries are expensive, so correctly sizing your battery bank is crucial. You should have a minimal battery storage of two-thirds of your daily consumption to supply power at night.

Power Inverters

Imagine your solar panels as diligent workers, gathering sunlight and transforming it into DC electricity. Inverters are brilliant managers who convert this raw energy into AC electricity. This then powers everything in your home, from the refrigerator in your kitchen to the lamp in your living room. Understanding the types of inverters, how to size them correctly, their efficiency, and where to place them is crucial for optimizing your off-grid solar power system.

Types of Inverters

In the world of solar power, inverters come in several flavors, each suited to particular needs and setups:

- *String Inverters*: These are the traditional choice for connecting a series (or string) of solar panels together. The DC electricity from the panels is sent as a collective group to the inverter, which then converts it into AC electricity. They're cost-effective and straightforward but work best when solar panels receive uniform sunlight. If one panel underperforms—say, due to shading—it can affect the output of the entire string.

- *Microinverters*: Imagine giving each solar panel its own mini inverter. That's what microinverters do. They're attached directly to individual panels, converting DC to AC immediately. This setup means that each panel operates independently, so if one panel's performance dips, the others are unaffected. While microinverters are more expensive upfront, they maximize overall system efficiency and allow for easier expansion down the line.

- *Hybrid Inverters*: These are multitaskers that combine the functions of a standard inverter with a battery charger. They can manage input from the solar panels, charge the battery system, and even connect to the grid if necessary. Hybrid inverters are ideal for systems designed for energy storage or those that want the flexibility to add storage later.

Power Inverter

Size Your Inverter

Matching your inverter's capacity to your solar panel output ensures you're efficiently converting as much solar energy as possible. Here's the lowdown:

- Calculate your solar array's total wattage. This is your starting point.

- Choose an inverter that matches or slightly exceeds your panel's peak wattage. For example, if your panels can generate 5,000 W at peak performance, choose an inverter rated for at least 5,000 W.

- Consider future expansions. If you plan to add more panels, opting for a larger inverter now could save you from needing an upgrade later.

Efficiency Factors

An inverter's efficiency tells you how much DC electricity is successfully converted into usable AC electricity. Several factors influence this:

- *Temperature*: Inverters prefer the Goldilocks zone—not too hot or cold. Extreme temperatures can reduce efficiency, so climate control or shading can help maintain optimal performance.

- *Loading*: Running an inverter close to its maximum capacity can optimize efficiency. An inverter too large for your system might run less efficiently at lower loads.

- *Quality*: Not all inverters are created equal. Higher-quality models often boast better efficiency, producing more usable electricity and less waste.

To get the most from your inverter, pay attention to these factors and consider them when selecting and setting up your system.

Installation and Placement

Where and how you install your inverter can significantly impact its performance and lifespan. Here are some best practices:

- *Ventilation*: Keep it cool. Inverters generate heat, and excessive heat can lead to inefficiency or damage. Ensure your installation spot is well-ventilated, and if you live in a hot climate, consider adding an active cooling method.

- *Accessibility*: Place inverters where they can be easily accessed for maintenance or monitoring. While it might be tempting to tuck them out of sight, being able to check on and service your inverter without a hassle is a boon.

- *Distance*: Keep inverters close to your battery bank and solar panels to minimize energy loss through long cables. The shorter the distance, the less power is lost in transmission.

- *Protection*: Protect your inverter from the elements—direct sunlight, rain, and snow can lead to problems. An indoor installation or a protective outdoor enclosure works wonders.

Inverters are more than just middlemen in the solar power process. They're key players in ensuring your off-grid system runs smoothly, efficiently converting and delivering power where and when needed. By choosing the right type, sizing it correctly, focusing on efficiency, and installing it thoughtfully, you set the stage for a solar power system that meets your needs today and is ready to grow with you into the future.

Connect to the Grid

For grid-tied systems, partnering with your utility company from the get-go is vital. Here's a simplified process:

- *Net Metering Agreement*: This pact with your utility company allows you to feed excess power back into the grid in exchange for credits. It requires an application and approval process that varies by location.

- *Grid Interconnection*: Once you have your net metering agreement, a professional must connect your solar system to the grid. This often involves installing a bi-directional meter installation that tracks the energy you consume and the excess you generate.

- *Inspection and Approval*: Before your system goes live, a final inspection will be done by the utility company or a city inspector to ensure everything is up to code and safely connected. Passing this inspection is crucial to get the green light to activate your system.

Permits and Regulations

Navigating legal and permit requirements for setting up an off-grid system can be a frustrating experience. However, it is a necessary step to ensure that your system is fully compliant with the relevant regulations. The last thing you want is to find out down the line that you need to make costly adjustments. This is discussed in more detail in *Chapter 10: Navigate the Legal Landscape*.

Consider Using Solar Panel Kits

A solar panel kit is a package that contains all the essential components required to set up a solar power system. Typically, it includes solar panels, mounting hardware, inverters, wiring, and sometimes battery storage. These kits are available in various sizes and configurations to meet energy needs and installation requirements.

The benefits of using a solar panel kit are numerous:

- *Convenience*: Solar panel kits are designed to simplify the process of going solar by bundling all the essential components into one package. This convenience saves time and effort by eliminating the need to source individual components separately.

- *Easier Installation*: Solar panel kits come with detailed instructions that guide users through the installation process step-by-step. This makes it easier for DIY enthusiasts to set up their solar power systems without needing specialized skills or professional assistance.

- *Cost-Effectiveness*: Purchasing a solar panel kit is often more cost-effective than buying individual components separately. By bundling everything together, manufacturers can offer kits at a lower price, making solar energy more accessible and affordable for homeowners.

- *Scalability*: Many solar panel kits are modular and scalable, allowing users to expand their solar power system over time as their energy needs grow. This flexibility enables homeowners to start small and add more solar panels as their budget allows or as their energy consumption increases.

- *Warranty Coverage*: Solar panel kits from trustworthy manufacturers generally include warranty coverage for their components, giving users assurance and protection against any defects or malfunctions. This warranty coverage ensures that users can rely on their solar power system's performance and receive support as needed.

Overall, solar panel kits offer a convenient, cost-effective, and user-friendly solution for homeowners looking to harness the power of solar energy and reap the benefits of renewable power generation. Whether you're a DIY enthusiast or just looking to save money on your energy bills, a solar panel kit can be an excellent investment for your home.

Professional Consultation

When tackling off-grid projects like setting up a solar power system, it's all about finding that sweet spot between DIY enthusiasm and knowing when to call in the pros. Sure, it's fantastic to dive into a project headfirst, but let's be honest: some things need specialized skills or are downright safety hazards. If you've got the budget, there's no shame in reaching out for professional assistance, especially if your skill set isn't entirely up to par.

Whether electrical work needs a pro's touch or structural tweaks that require some finesse, investing in expert help ensures your project gets done safely and up to code. So, don't sweat it if you're not a DIY guru—there's no shame in knowing when to pass the torch to someone with the know-how.

Your Solar System Design

Bring together the calculations and knowledge from this chapter to summarize your solar system design. Use this table to input your data for a quick reference:

Component	Input
Energy Needs (kW) Use the results from your energy audit in *Chapter 1*	
Solar Panels • Type • Quantity	
Mounting • Location • Hardware • Orientation & Angle	
Charge Controller • Type (PWM or MPPT) • Capacity (kW) • Settings	
Battery • Type • Capacity (kWh) • Rated Power Output (kW)	
Inverter • Type • Capacity (kWh) • Placement • Grid-Tied or Off-Grid	
Compliance • Local Permits • Documentation	

Wrapping It Up...

We have comprehensively analyzed the various components of a solar power system to determine the best solution for your needs. This analysis included an evaluation of different types of solar panels and their respective efficiencies and an in-depth examination of the complexities associated with inverters and controllers, which are crucial for converting and managing solar energy. We also considered available mounting solutions and made recommendations for battery selection.

In the next chapter, we will provide a step-by-step guide to help you install your solar power system, allowing you to harness the sun's energy and reduce your carbon footprint.

4

STEP-BY-STEP SOLAR DIY

Imagine a day where your actions do not harm the planet, and the only trace you leave behind is your footprints on the sand. You can start this new day by harnessing the sun's power and converting it into valuable energy right on your doorstep. Solar installation is not just about placing solar panels on your roof and hoping for the best. It requires a well-thought-out plan that transforms your home into a sustainable powerhouse.

As you progress, remember to follow the manufacturer's guide when installing a solar power system. The guide helps you avoid mistakes, keep your system running smoothly, and ensure safety. It also helps you make warranty claims. Ignoring the guide can lead to issues. So, treat the manual like your best friend – it will guide you towards a successful and hassle-free installation.

Step 1: Site Assessment and Positioning

A top-notch solar installation begins with a keen eye for detail and a solid understanding of your environment. It's like finding the perfect spot in your garden where the sun lingers long enough to help your tomatoes thrive.

Analyze Sun Paths

Understanding how the sun travels across your property is crucial. Observing the sun's path helps you pinpoint areas that receive uninterrupted sunlight for extended periods. Tracking these patterns at different times of the year is beneficial since the sun's trajectory changes with seasons. Simple tools like a compass, Google Earth, or sun path apps can provide insights into the sun's movement, helping you align your panels for the best exposure. This tech-savvy approach and old-fashioned observation ensure you pinpoint the sweet spots that promise maximum solar gain.

Positioning

Positioning solar panels to capture the most sunlight involves more than just facing them towards the sky. It's a fine-tuned process that blends science with landscaping artistry. Optimizing panel placement ensures your solar setup performs at its peak, turning every ray of sunshine into valuable electricity. Here, we navigate through practical steps to ensure your panels bask in the optimal sunlight:

- *Sun Exposure*: Look for spots that get direct sunlight for most of the day. Remember, even a tiny patch of shade can significantly reduce your system's efficiency.

- *Roof Condition and Orientation*: If you're eyeing the roof, its condition, angle, and direction are critical. South-facing roofs in the Northern Hemisphere catch more sun but don't discount east or west-facing slopes.

- *Ground Conditions*: Considering a ground mount? Then, the terrain's slope, soil type, and stability come into play. Flat, stable land free of obstructions is ideal.

Mitigate Shading

Shadows from trees, buildings, and other structures can dramatically reduce your solar panels' output. Even a tiny amount of shade on one panel can affect the entire system's performance. To mitigate shading:

- *Trim Foliage*: Regularly trimming trees and bushes can prevent them from casting shadows on your panels.

- *Strategic Placement*: Install panels in locations that remain clear of shadows throughout the day. Sometimes, there are more obvious ones than the ideal spot.

- *Microinverters*: Using microinverters can help lessen the impact of shading. Since each panel operates independently, the shaded panel doesn't drag down the performance of the entire array.

Seasonal Adjustments

The sun doesn't stand still, and neither should your solar panels. Adjustable mounting systems, which allow you to change the tilt angle of your panels with the seasons, can significantly boost your system's efficiency. During summer, a shallower angle captures the high sun, while a steeper tilt in winter catches the lower sun. Although adjusting the panels requires a bit of effort, the increase in solar harvest is often worth the trouble.

Step 2: Your System Design

After finding the perfect location, the next step is to identify a solar power system that caters to your energy needs and site specifics. It's like solving a puzzle where every component must fit perfectly. The previous chapter discussed the components required for a successful implementation. You can use the results you entered in *Chapter 3 - Your Solar System Design Parameters.*

With this design blueprint, you're not just preparing for installation; you're also laying the foundation for a system that's efficient, reliable, and customized to your lifestyle.

Step 3: Tools and Equipment

Finally, getting your toolkit ready is not just about preparation; it's about empowerment. Having the right tools means you're prepared to tackle the installation confidently. Here's an essential checklist to get you started:

- *Safety Gear*: Safety glasses, gloves, and sturdy footwear are nonnegotiable. Protecting yourself from potential mishaps is priority number one.

- *Hand Tools*: A set of screwdrivers, wrenches, pliers, and a hammer will cover most of your needs. Think of them as the trusty sidekicks in your solar installation adventure.

- *Power Tools*: A drill and a saw might be helpful, especially for custom mounts or retrofitting areas for your panels.

- *Measuring Instruments*: A tape measure, level, and possibly a laser distance measure will ensure your panels are positioned precisely.

This toolkit, both literal and metaphorical, equips you for the physical act of installation and the journey of transforming your home into a bastion of sustainability. With every tool you wield and every panel you mount, you're a step closer to a future where clean, renewable energy is not a luxury but a given.

Step 4: Electrical Safety

In DIY solar power, safely weaving through the web of electrical components is as critical as the morning sun rising in the east. The electrifying journey of turning photons into usable power teems with potential hazards. However, with the right know-how and gear, it's a path paved with rewards. Let's illuminate the essentials of electrical safety, ensuring your solar installation radiates success without any shocking surprises.

If Unsure...

Tackling a DIY solar panel installation is quite the adventure, full of potential and excitement! However, when it comes to the electrical parts, things can get a bit tricky. It's really important to remember that dealing with wires and electrical setups requires specific know-how to avoid any mishaps. We don't want any accidents or, worse, fires to dampen your green energy journey. So, if you are unsure, reach out to a professional electrician or solar expert. They can help make sure everything's set up safely and efficiently, keeping you and your home safe.

Understand Electrical Basics

Grasping the core principles of electricity forms the bedrock of any safe solar installation. Electricity, in its simplest terms, flows like water, with current (measured in amperes or amps) representing the flow rate and voltage (measured in volts) symbolizing the pressure that drives this flow. Grounding, or providing a path for electrical current to safely return to the earth in case of a fault, is the safety net that keeps the system from becoming a hazard. Here's what you need to keep in mind:

- *Current*: Solar panels and batteries can generate significant currents. Always respect this power; even a low-voltage system can produce a dangerous current under certain conditions.

- *Voltage*: While solar panels operate at DC voltages that are generally considered safe, combining multiple panels can create higher, more hazardous voltages.

- *Grounding*: Proper grounding of your solar system is non negotiable. It minimizes the risk of electric shock by providing a safe path for fault currents, directing them away from you and into the earth.

Personal Protective Equipment (PPE)

Donning the proper armor, or personal protective equipment (PPE), is your first defense against potential electrical mishaps. This gear is not about fashion; it's about creating a barrier between you and the electrical forces you work with. Make sure you have:

- *Gloves*: Insulated gloves are a must as they shield your hands from electric shocks and burns. They should be dry and free from tears or punctures.

- *Eyewear*: Safety glasses protect your eyes from sparks, debris, and UV radiation, mainly when you're working outdoors or cutting materials.

- *Footwear*: Electrically insulated boots ground you, preventing electric current from passing through your body if you encounter a live wire.

Safe Installation Practices

Navigating the electrical landscape of your solar setup safely requires more than just the right gear; it demands a meticulous approach to handling and installing components. Here are some golden rules:

- *Power Off*: Always ensure the power is off before tinkering with any part of the system. This includes disconnecting batteries and switching off breakers.

- *No Metal Jewelry*: Rings, bracelets, and watches can become conductive hazards. Leave them off when you're working.

- *Use Insulated Tools*: Tools with insulated handles add an extra layer of protection, reducing the risk of accidental shocks when working with live components.

- *Proper Wiring Techniques*: Secure connections prevent loose wires, which can cause shorts and sparks. Follow manufacturer guidelines for stripping, connecting, and securing all wiring.

Emergency Preparedness

Even with the best precautions, things can go awry. Being prepared for electrical emergencies is part of being a responsible solar installer. Here are some ways you can stay ready:

- *Know Your Breakers*: Familiarize yourself with the electrical panel and know which breakers control the power to your work area. In an emergency, shutting off the correct breaker can prevent further hazards.

- *Fire Extinguisher*: Keep a class C fire extinguisher nearby. Electrical fires require specific extinguishing agents, and water can worsen the situation.

- *First Aid Kit*: Have a first aid kit tailored for electrical burns and injuries. Knowing basic first aid procedures can make a significant difference in the outcome of an accident.

- *Emergency Plan*: Have a plan in place for electrical emergencies. This includes knowing how to contact emergency services and having clear access to exits.

When installing solar panels, it's essential to prioritize safety. By familiarizing yourself with electrical safety, wearing appropriate PPE, following safe installation practices, and preparing for emergencies, you can ensure a successful and secure installation process. This generates power for your home and demonstrates your respect for the powerful forces at play in the process.

Step 5: Solar Panel Installation

Installing solar panels, whether on the roof or the ground, begins with a clear plan and the right tools. It's akin to setting up a complex Lego set—follow the instructions, and you'll end up with a functional, efficient solar power setup.

Mount the Panels

First up, let's talk about getting those panels up and ready:

- *Roof Mounts*: Start by ensuring your roof can handle the weight. Panels weigh, on average, 40 lbs. Use mounting brackets specifically designed for solar installations. These brackets are typically secured directly into the roof rafters for stability. Position your panels where they will get maximum sunlight, avoiding commonly shaded areas. Use a drill to secure the brackets, then attach the mounting rails. The panels then lock into place on these rails.

- *Ground Mounts*: You must pour concrete foundations to hold the mounting structure. Once the concrete sets, assemble the metal framing that will support your panels. This structure must be angled correctly to optimize sun exposure based on your specific latitude.

- *Pole Mounts*: To adjust your solar panels for maximum sun exposure, choose a sturdy pole mount made of durable materials like galvanized steel or aluminum. Secure the pole into the ground using concrete for stability. Attach a mounting bracket to the pole and ensure the mount can be tilted and rotated for optimal alignment with the sun's path.

In these cases, ensure your panels are angled in such a way that they maximize sun exposure throughout the day. An inclinometer can be pretty handy for this task.

Step 6: Wiring Your System

Connecting your panels to inverters and batteries requires attention to detail. You're dealing with DC electricity generated by the panels, which will be converted to AC by the inverters for use in your home or to feed back into the grid.

Types of Cables

- *PV Cables*: PV cables are specifically designed for the outdoor environment and direct exposure to sunlight. These cables are UV resistant and have a high temperature and weather resistance, making them ideal for connecting solar panels.

- *AC Cables*: These cables are used to connect the inverter to the home electrical system. They are typically standard household electrical cables but must be sized correctly for the current they will carry.

- *Battery Cables*: These are heavy-duty cables designed to handle the high current flow between the battery bank and the inverter or charge controller. They are usually thicker to minimize voltage drop and heat buildup.

Cable Specifications

- *Voltage Rating*: The cable must be suitable for the system's maximum voltage. For most residential systems, cables rated for 600 V are common, but systems with higher voltage may require cables rated at 1000 V or higher.

- *Current Carrying Capacity*: The cable must be able to handle the maximum current expected from the panels, plus a safety margin. This is often calculated based on the short-circuit current of the solar panels.

- *Size (Gauge)*: The size of the cable is critical and is based on the distance of the run (from the panels to the inverter and from the inverter to the grid connection) and the current it needs to carry. Using a cable that's too small can lead to significant power losses and even pose a fire hazard. The American Wire Gauge (AWG) system is commonly used in the U.S.

Types of Connectors

- *MC4 Connectors*: These are the most common type of connectors used in solar PV systems. They are designed to connect solar panels in series or parallel, are waterproof, and can be easily disconnected and reconnected. MC4 stands for the manufacturer Multi-Contact and a 4 mm² contact assembly pin.

- *Anderson Connectors*: These are sometimes used for connecting to battery banks or other types of equipment. They are known for being robust and making a solid connection.

- *Terminal Blocks*: These are used for connections inside combiner boxes or for terminating cables in the inverter.

Safety and Compatibility

- *Compatibility*: Ensure that the connectors are compatible with the cables and components they will connect with, including the correct gender for MC4 connectors and matching terminal sizes for blocks.

- *UV and Weather Resistance*: Ensure that any cables and connectors that are used outside are rated for UV exposure and weather conditions.

- *Electrical Codes and Standards*: Follow the National Electrical Code (NEC) or equivalent standards in your region for solar installations, including requirements for overcurrent protection and grounding.

- *Labeling*: Label all components and wiring for future reference and maintenance.

Grounding

Separate grounding wires may be needed to ground the solar panel frames and other metallic parts of the installation to protect against lightning strikes and reduce electrical noise.

Connect Your System

When planning your DIY solar system, carefully calculate the cable sizes using appropriate formulas or online calculators, considering the total amperage, voltage drop, and distance of cable runs. Incorrect cable sizing can lead to inefficient system performance or even safety hazards. It's also wise to purchase a little extra cable and a few additional connectors to allow for mistakes or future expansions.

Follow these steps when connecting your system:

- *Start by Connecting the Panels*: Use appropriate gauge solar cables to connect your panels in series or parallel, depending on your system's voltage requirements. Ensure all connections are tight and secure to minimize the risk of electrical loss.

- *Inverter Connection*: From your solar array, the next stop is the inverter. This might be a central inverter for the whole system or individual microinverters behind each panel. Ensure the inverter is installed near the main panel to reduce voltage drop.

- *Battery Bank*: Connect the output from your inverter to your battery bank. This often involves heavy-duty cables capable of handling the system's total amperage. A charge controller between the panels and batteries will regulate the charging process, ensuring batteries are neither overcharged nor overly depleted.

Safety is paramount. Always ensure power is off when making electrical connections and use insulated tools to prevent accidental shocks.

Step 7: Test and Commission

Before you start enjoying your new solar power, a thorough testing phase ensures everything works as it should. The testing stage involves:

- *Visual Inspection*: Check every connection, ensuring

they're tight and correctly wired. Look for any signs of damage to panels, wires, or the inverter.

- *Voltage Testing*: Using a multimeter, verify that your panels are generating the expected voltage. Check at various points–panel output, inverter input, and inverter output–to ensure no unexpected drops.

- *Inverter Testing*: Most inverters have a testing mode you can initiate. This checks the inverter's ability to convert DC to AC efficiently and to shut down in case of a grid failure, a safety requirement known as anti-islanding.

- *System Monitoring Setup*: Finally, configure any system monitoring tools you have. These could be apps or online platforms that track your system's performance. They're invaluable for catching issues early and optimizing your setup.

This meticulous, step-by-step approach ensures your DIY solar installation not only meets your energy needs but does so safely and efficiently. With every panel mounted, every wire connected, and every test conducted, you're laying the groundwork for a sustainable, energy-independent future.

Wrapping It Up...

This DIY installation guide covered the essential tools and safety equipment needed, highlighted the importance of electrical safety, and detailed the installation process, including panel mounting, system wiring, and grid connection. It emphasized positioning panels for maximum efficiency, adjusting for seasonal sun paths, and mitigating shade. Regular maintenance, such as cleaning panels and checking connections, is also stressed to ensure ongoing efficiency. This chapter serves as a practical roadmap for anyone looking to harness solar power, focusing on safety, efficiency, and sustainability.

Moving forward, we'll switch gears and look into the world of wind energy. Just as we've explored how to capture and optimize solar power, we'll uncover how harnessing the wind can complement your quest for sustainable, off-grid living.

5

RIDE THE WIND

Picture yourself standing on a hill, feeling the brisk and refreshing wind against your skin. This force of nature has been powering ships, grinding grain, and even flying kites for centuries. Today, it is at the forefront of the renewable energy revolution, helping us turn our desire for a cleaner planet into a tangible reality. Wind power is not just about gigantic blades spinning in the breeze; it is a symphony of engineering and environmental synergy that brings sustainable energy to our doorsteps.

The Basics

Wind turbine technology uses the power of the wind to generate electricity, providing a clean, sustainable, and cost-effective energy source. A wind turbine comprises large blades positioned on a tower that rotate when the wind blows. This movement powers a generator inside the turbine, which converts the wind's kinetic energy into electrical energy. Wind turbines come in different sizes, from small systems for individual home use to large turbines grouped in wind farms for utility-scale power generation. The blades' shape, angle to the wind, and speed are all carefully designed to capture as much of the wind's energy as possible.

Wind Turbines Types

When it comes to harnessing the wind, there are two main actors on the stage: horizontal-axis wind turbines (HAWTs) and vertical-axis wind turbines (VAWTs).

- *HAWTs*: These are the ones you likely picture when you think of wind turbines. They have a tall tower with blades that rotate around a horizontal axis. They must be facing the wind, so many modern HAWTs have a built-in yaw mechanism that automatically aligns them with the wind direction. They're efficient and have high power outputs, making them popular for utility-scale wind farms.

- *VAWTs*: Less common but increasingly attractive for residential and urban settings, VAWTs have a vertical rotor shaft. One of their key benefits is they don't need to be pointed into the wind to be effective. This makes them versatile and suitable for areas where wind directions frequently change. They're often more compact than HAWTs, offering the potential for rooftop or ground-level installations closer to where the power will be used.

Horizontal-Axis Wind Turbine (HAWT)

Vertical-Axis Wind Turbine (VAWT)

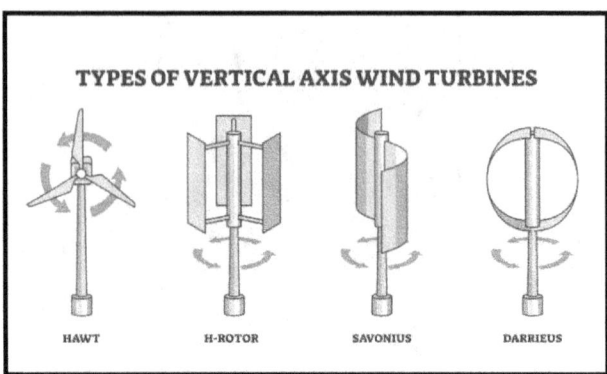

Types of Vertical-Axis Wind Turbines

Wind Turbine Components

Diving deeper into the anatomy of a wind turbine reveals a complex machine made up of several key components:

- *Rotor and Blades*: The most visible parts of the turbine, the blades and rotor, capture wind energy and convert it into rotational motion.

- *Nacelle*: Sitting atop the tower, the nacelle houses the generator, gearbox (in some turbines), and other mechanical components. It's essentially the brain of the operation.

- *Tower*: The tall structure that supports the nacelle and rotor, raising them above obstacles on the ground to better catch the wind.

- *Foundation*: The unsung hero of wind turbine construction, the foundation anchors the structure to the ground, ensuring stability against wind and gravity forces.

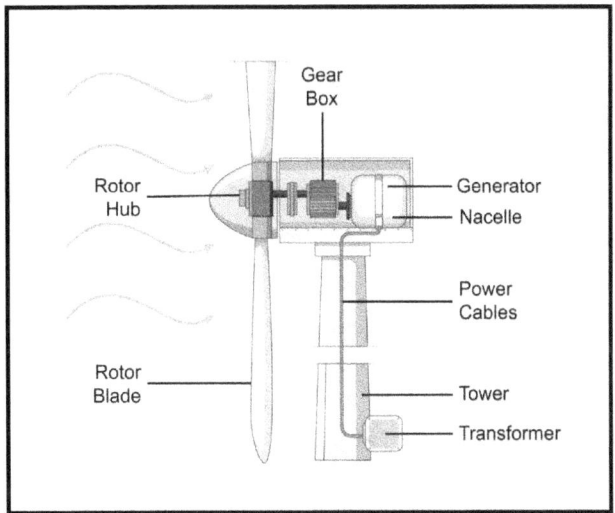

Wind Turbine System Components

Wind Turbine Selection

Choosing the suitable wind turbine for your off-grid setup or backyard is much like finding the perfect pair of shoes; it's all about fit, comfort, and environmental suitability. Whether for reducing your carbon footprint or cutting down on electricity bills, the journey starts with understanding your local wind resources, matching turbine size to your energy needs, and considering many logistical factors.

Assess Wind Resources

Before even thinking about the types of turbines, it's vital to know what you're working with–the wind itself. Wind maps are your first port of call, offering a bird's eye view of wind speeds in your area. These maps, available through government and private entities, provide average wind speeds at different heights, giving you a preliminary idea of the wind's potential to power a turbine.

However, local factors such as terrain, nearby bodies of water, and even artificial structures can dramatically affect wind flow. This is where site-specific wind speed measurements come into play. Using an anemometer, a device designed to measure wind speed, you can collect data at the exact location and height where you plan to install your turbine. A detailed assessment over a year gives you the best picture, capturing seasonal variations that could impact your turbine's performance.

To ensure your location can sustainably power a turbine, wind speed should have an average wind speed of around 9 mph.

Turbine Size and Power Output

With a clear understanding of your wind resource, the next step is aligning your energy needs with the right turbine size and power output. It's a balancing act:

- *Energy Needs*: Start by calculating your average energy consumption. Look at your utility bills or use the results from your DIY energy audit in *Chapter 1*.

- *Rated Power Output*: This is the maximum power output the turbine can produce under ideal conditions, typically measured in kilowatts. This rating is achieved at a specific wind speed, known as the rated wind speed, which is usually around 20 to 25 mph for most commercial turbines.

- *Cut-In Wind Speed*: This is the minimum wind speed at which the turbine starts generating electricity, usually between 6 to 9 mph.

- *Cut-Out Wind Speed*: This is the wind speed at which the turbine automatically shuts down to avoid damage, typically around 55 mph.

Understanding a turbine's capacity is crucial for evaluating its potential energy output. The rated capacity represents the maximum power the turbine can produce under ideal conditions. However, the actual energy a turbine generates over time – its capacity factor – is influenced by how consistently the wind blows at a site. For instance, a 10 kW turbine might only produce around 30-40% of that, depending on the site's wind resource. It's a reminder that while wind power is abundant and renewable, its delivery fluctuates as the wind.

Using our previous example for solar power requirements from *Chapter 3 - Power Requirements*, a household needing 15 kWh power consumption can be met by a 3 kW wind turbine operating at its rated wind speed for 5 hours to produce 15 kWh. This example assumes there is off-grid battery storage.

Turbine Power (kW) = Daily Energy Use (kWh) / hours (at rating wind speed)

= 15 kWh / 5 hrs

= 3 kW

So remember, a turbine that's too small won't meet your energy needs, while one that's too large might not be cost-effective. Consider your peak energy requirements and choose a turbine that meets or slightly exceeds this demand under average local wind conditions.

Power Output: AC or DC

Wind turbines convert the kinetic energy from the wind into electrical energy, which can be output as either AC or DC. AC is much preferred for wind turbines due to compatibility and transmission advantages, while DC has niche applications, particularly where storage efficiency is paramount.

For an AC wind turbine, a rectifier is needed to convert to DC for battery charging. Similar to Solar Power Systems, an inverter is required to convert the DC power from the battery into AC to power house appliances.

Turbine Placement

Placing your wind turbine is as much an art as a science. Three critical factors come into play:

- *Height*: Wind speeds increase with elevation from the ground due to the reduced effect of drag from the earth's surface. A taller tower can access faster, less turbulent wind, improving turbine efficiency. However, tower height can be limited by local zoning regulations or logistical challenges.

- *Distance from Obstacles*: Trees, buildings, and other structures can create wind shadows and turbulence, reducing the efficiency of your turbine. As a rule of thumb, a turbine should be installed at least 30 ft above anything within a 500-ft radius.

- *Zoning Regulations*: Your local planning office can be a goldmine of information on what's allowed and what's not. Some areas have restrictions on tower height, noise levels, or even the turbine's appearance. Getting familiar with these regulations early on can save you a lot of headaches down the line.

Rectifiers

Rectifiers in wind turbines convert AC produced by the wind turbine generator into DC power. This conversion is essential for charging batteries, powering DC systems, or for further inversion back to AC, but at a controlled frequency and voltage suitable for grid integration or direct use.

- *Silicon-Controlled Rectifier (SCR)*: SCR systems are robust, offering precise control over the output voltage, making them suitable for applications requiring high reliability and efficiency, such as in large-scale wind farms.

- *Diode Bridge Rectifier*: Commonly used in smaller wind turbines, diode bridge rectifiers are simpler and cost-effective. They are best suited for applications where the power demand is relatively low and constant, such as in stand-alone wind systems powering remote locations.

Power Inverters

Inverters take DC power from your batteries and convert it to AC for use in your home. This AC can also be matched with the grid's frequency and voltage if you're feeding back into the grid. There are three types of inverters you should be aware of:

- *Stand-Alone Inverters*: These inverters are used in off-grid systems, where the wind turbine's energy is not intended for grid export but for local use. They are designed to meet the specific demands of the application, whether it be household appliances, small workshops, or remote communication stations.

- *Grid-Tied Inverters*: Specifically designed to synchronize with the electrical grid, these inverters ensure that the energy produced by wind turbines can be directly supplied to the grid. They incorporate features such as anti-islanding protection, MPPT, and compliance with grid codes.

- *Battery-Based Inverters*: These inverters are pivotal in systems where energy storage is essential. They manage the flow of energy to and from batteries, ensuring that excess energy produced can be stored and used when wind conditions are not favorable.

The application of specific inverters and rectifiers in wind turbines is governed by several factors, including the scale of the wind installation, its location, and the intended use of the generated power. Large-scale wind farms typically lean towards SCR rectifiers and grid-tied inverters for their efficiency and grid compatibility, while smaller, isolated applications might prioritize diode bridge rectifiers and stand-alone inverters for their simplicity and cost-effectiveness.

Charge Controllers

A charge controller, sometimes called a charge regulator, is an essential part of almost any wind turbine system, including a battery bank. Its primary function is to regulate the batteries' charging, preventing overcharging and deep discharge, which can significantly harm battery health. By doing so, it extends the batteries' life and maintains the system's overall efficiency.

Types of Charge Controllers

There are two main types of charge controllers used in wind turbine systems:

- *PWM Controllers*: These are the more traditional controllers that regulate battery charging by switching the charging current on and off quickly. While simpler and usually more affordable, they might only sometimes offer the best efficiency in managing the erratic power output from wind turbines.

- *MPPT Controllers*: MPPT controllers are more sophisticated and can adjust their input to capture the maximum power from the wind turbine at any point in time. They convert excess voltage into amperage, which results in a more efficient charge to the battery. This type is particularly beneficial for wind systems due to the variable nature of wind speed and power generation.

Select a Charge Controller

When choosing a charge controller for a DIY wind turbine system, several factors need to be considered to ensure optimal performance:

- *Compatibility With Battery Type*: Ensure the charge controller is compatible with the type of batteries you are using (e.g., lead-acid, lithium-ion).

- *System Voltage*: The controller must match the system's voltage (12 V, 24 V, 48 V, etc.).

- *Current Capacity*: It should be able to handle the maximum current your wind turbine can produce.

- *Features and Safety Measures*: Look for additional features such as overcurrent protection, lightning protection, and reverse polarity protection. Some controllers also offer load control and display panels for monitoring system performance.

Integrate Charge Controllers

Integrating a charge controller requires a clear understanding of the system's electrical architecture in a DIY wind turbine setup. As most wind turbines generate AC power, a rectifier converts to DC power, which then charges the batteries via a charge controller.

Proper installation ensures that the wind turbine's variable power output is safely managed, optimizing battery charging and preventing potential damage to the system. It's also crucial to ensure the wiring and connections are secure and comply with electrical standards to avoid any safety hazards.

Battery Storage

Storing today's gusts for the calm of tomorrow is where battery storage shines in a wind energy setup. Batteries act as a bank, holding on to the energy generated by your turbine until it's time to light up your home, power your devices, or keep your refrigerator running. With the right size battery bank, you'll have a reliable reservoir of energy ready to meet your needs, rain or shine.

Integration With Other Energy Sources

The key to a resilient off-grid system lies in diversification. Just as a well-rounded diet combines different food groups, a robust energy system integrates multiple power sources. Here's how wind power can join forces with solar and other renewables:

- *Complementary Cycles*: Wind and solar resources often follow opposite patterns. Sunny days might not be windy, whereas windy conditions can occur on cloudy days. You can tap into nature's rhythm by combining wind turbines with solar panels, ensuring a more constant energy supply.

- *Hybrid System Controllers*: These smart devices can manage inputs from wind turbines and solar panels, charging your battery bank with whichever source is available or both simultaneously if conditions allow.

- *Energy Management Software*: Modern systems come equipped with software that decides the most efficient source to draw from at any given time, storing excess energy for later use. It's like having a financial planner for your power supply, always ensuring the best return on your energy investment.

Blending wind with other renewable sources maximizes your energy yield and adds an extra layer of security, ensuring you have power when you need it most.

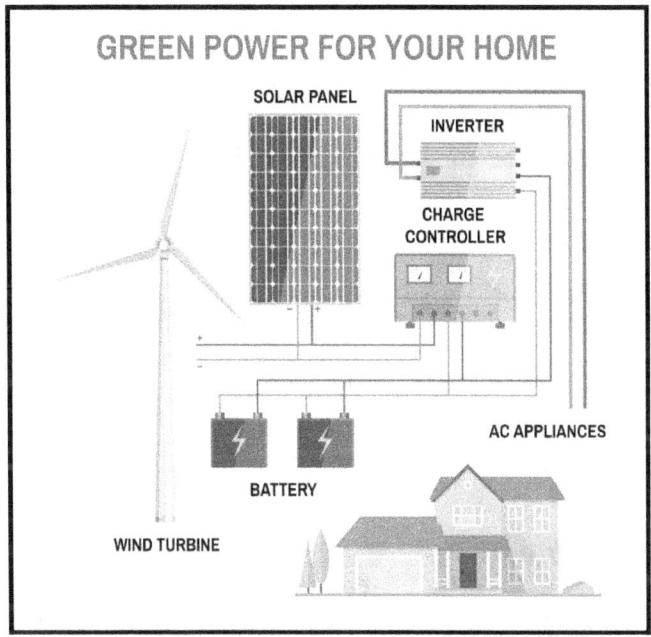

Hybrid Power System

Connect to the Grid

It's important to team up with your utility company early on if you plan to install a grid-tied solar system. Here's a simplified process to follow:

- *Net Metering Agreement*: You'll need to sign a legal agreement with your utility company that allows you to send excess energy back to the grid in exchange for credits. The application and approval process for this agreement may vary depending on your location.

- *Grid Interconnection*: Once you have your net metering agreement, a professional will connect your solar system to the grid. This typically involves installing a bi-directional meter that can track the energy you consume and the excess energy you generate.

- *Inspection and Approval*: Before your system can go live, it must pass a final inspection by the utility company or a city inspector to ensure everything is safely connected and up to code. Passing this inspection is crucial to activating your system.

Permits and Regulations

Setting up an off-grid system requires navigating legal and permit requirements to be fully compliant. *Chapter 10: Navigate the Legal Landscape* provides the necessary details for the steps you need to follow to meet these requirements.

Wind Turbine Kits

Like solar panel kits, wind turbine kits are designed to make installing a wind turbine more accessible, cost-effective, and satisfying for individuals looking to contribute to sustainable energy initiatives or reduce their electricity bills. Here are some of the key advantages:

- *Cost Savings*: Kits are generally more cost-effective than buying individual components separately.

- *Easier Installation*: The kits come with detailed, step-by-step instructions that allow DIY enthusiasts to set up their systems without professional assistance or specialized skills.

- *Customization and Scalability*: Wind turbine kits offer customization choices based on power output, turbine size, tower height, and critical factors that affect performance, ensuring maximum efficiency. Starting with small-scale installations allows one to increase energy independence by adding more turbines over time.

Professional Consultation

When tackling off-grid projects, balancing DIY enthusiasm with seeking professional help is crucial. Some tasks require specialized skills or pose safety risks. If you have the budget, contacting experts for help is okay. Whether electrical work or structural adjustments, investing in expert assistance ensures a safe and code-compliant project.

Your Wind Turbine System Design

Bring together the calculations and knowledge from this chapter to summarize your wind turbine system design. Use this table to input your data for a quick reference:

Component	Input
Energy Needs (kW) Use the results from your energy audit in *Chapter 1*	
Wind Turbine • Type (HAWTs or VAWTs) • Capacity (kW) • Rated Wind Speed (mph)	
Turbine Placement • Location • Height	
Rectifier (AC Turbine) • Type • Capacity	
Charge Controller • Type (PWM or MPPT) • Capacity (kW)	
Battery • Type • Capacity (kWh) • Rated Power Output (kW)	
Inverter • Type • Capacity (kWh) • Placement • Grid-Tied or Off-Grid	
Compliance • Local Permits • Documentation	

Wrapping It Up...

In this chapter, we've peeled back the layers of wind turbine technology, revealing the intricate dance between wind and machine that generates clean, renewable energy. From the aerodynamics of blades to the types of turbines dotting our landscapes and the critical components that make them tick, wind power stands as a testament to human ingenuity and our pursuit of sustainability. As we harness the wind, we're reminded of the power and potential of working in harmony with the natural world.

6

STEP-BY-STEP WIND TURBINE DIY

Installing a wind turbine is an adventure in self-sufficiency, blending technical know-how with a pioneering spirit. From breaking ground to the thrilling moment of watching your turbine catch its first gust, here's a detailed guide to bringing wind power to life at your doorstep.

When you're setting up a wind turbine system, sticking to the manufacturer's guide is important to get everything just right. These instructions are there to help you dodge common setup slip-ups, keep your turbine running smoothly, and make sure it's safe. Plus, if you ever need to claim the warranty, following the guide is your golden ticket. Ignoring these tips can lead to a bunch of issues down the road, and nobody wants that. So, treat the manuals like your best bud in this project and have a hassle-free installation.

Step 1: Site Preparation and Foundation

Before anything else, finding the right spot for your turbine sets the stage. A location with minimal obstructions from trees or buildings and ample wind flow is ideal. Once you've pinpointed the perfect site:

- *Clear the Area*: Ensure the ground is clear of debris, rocks, and vegetation. A flat surface makes the rest of the process smoother.

- *Mark the Layout*: Using stakes and string, outline the area where the foundation will be laid. This visual guide ensures accuracy in your work.

- *Dig the Foundation*: The depth and width depend on the turbine size and the manufacturer's recommendations. Generally, a deeper foundation provides better stability for taller turbines.

- *Pour the Concrete*: Mix and pour concrete into the foundation pit. Insert anchor bolts to secure the turbine tower into the wet concrete, following precise measurements from the turbine's installation manual. Allow it to cure fully, which can take several days.

Step 2: Your System Design

After identifying and preparing the location for your wind turbine, you can refer to *Your Wind Turbine System Design Parameters* from the previous chapter. This table contains the necessary information to design an efficient, reliable, and personalized system that will suit your lifestyle. You can use this as a blueprint to guide you through the entire design process.

Step 3: Assemble and Erect

With the foundation set, assembling your turbine is the next major step. This phase requires patience and precision:

- *Assemble on the Ground*: If possible, start assembling parts of the turbine on the ground. This might include attaching the blades to the rotor or assembling sections of the tower.

- *Raise the Tower*: Lifting the tower to its standing position is a critical step that often requires mechanical assistance, such as a crane or a winch, depending on the tower's height and weight. Safety is paramount here; ensure the area is clear and all helpers know their roles.

- *Secure the Tower*: Once upright, secure the tower to the foundation using the anchor bolts. Double-check all connections for tightness and alignment.

Step 4: Wiring Your System

Wiring a DIY wind turbine system requires careful planning and execution to ensure safety and efficiency. This guide assumes a basic setup including a wind turbine, rectifier (AC to DC power), charge controller, battery bank, inverter (DC to AC power), and necessary safety devices.

Safety First

Ensure all components are turned off or disconnected before beginning, and wear appropriate safety gear.

Just like the solar panel installation in *Chapter 4*, a DIY wind turbine installation is a bit of an adventure. However, it also has its potential dangers as you're dealing with electrical components.

If you are unsure, stay safe and avoid mishaps by reaching out to a professional electrician or wind turbine expert. They can help you set everything up safely and efficiently.

Types of Cables

- *PV Cables or Wind Turbine Cables*: These cables are designed to withstand outdoor environments, including UV exposure, temperature variations, and possibly harsh weather conditions. They're used to connect the wind turbine to the charge controller and other system components. They need to be durable, with a high enough voltage rating to handle the system's output, plus a safety margin.

- *Battery Cables*: These are heavy-duty cables designed to handle the high current flow between the battery bank and the inverter or charge controller. These are usually thicker to minimize voltage drop and heat buildup.

- *AC Cables*: If your system includes an inverter to convert DC to AC, you'll need suitable AC cables to connect the inverter to your home's electrical panel or the grid. The size and type depend on the inverter's output and local electrical codes.

Cable Specifications

- *Length and Size (Gauge)*: Calculate the necessary cable lengths carefully to minimize voltage drop, which can significantly reduce efficiency. Use online calculators or formulas to determine the appropriate wire gauge based on current, voltage, and length of run.

- *Voltage Rating*: The cable must be suitable for the system's maximum voltage. For most residential systems, cables rated for 600 V are common, but systems with higher voltage may require cables rated at 1000 V or higher.

- *Current Carrying Capacity*: The cable must be able to handle the maximum current expected from the panels, plus a safety margin. This is often calculated based on the short-circuit current of the solar panels.

- *Safety Standards*: Comply with local electrical codes and standards. This might involve using specific types of conduit or raceway for protection and ease of maintenance.

Selecting the correct cables and connectors is vital for the safety, performance, and durability of your DIY wind turbine system. Always prioritize quality and compatibility to ensure a reliable and efficient setup.

Types of Connectors

- *MC4 Connectors*: While more common in solar installations, MC4 connectors are also suitable for wind turbine connections due to their durability, weather resistance, and ability to handle high voltages and currents.

- *Ring Terminals*: These are used for secure connections to battery terminals and grounding points. They should be of appropriate size for the cable and terminal post.

- *Anderson Connectors*: These connectors are useful for modular systems or where quick disconnection and reconnection might be needed, especially between the turbine and the charge controller or battery bank.

- *Terminal Blocks*: These are used for connections inside combiner boxes or for terminating cables in the inverter.

- *Cable Glands*: While not connectors in the traditional sense, cable glands are essential for passing cables into enclosures (like the charge controller or battery box) while maintaining a weather-tight seal.

Always ensure that every connector you use fits securely with the components they will connect. This may require checking the manufacturer's specifications for each part of your wind turbine system.

Selecting the correct cables and connectors is vital for the safety, performance, and durability of your DIY wind turbine system. Always prioritize quality and compatibility to ensure a reliable and efficient setup.

Grounding

Always ground the wind turbine, tower, and electrical system components following local electrical codes. This step is crucial for safety, protecting against lightning strikes and electrical faults.

Connect Your System

Connecting your turbine to your home's power system is where you see your efforts start to pay off. Assuming we have a wind turbine that generates AC power, here's how to connect your system:

- *Wire the Turbine*: Connect the turbine to a rectifier to convert AC to DC power so the batteries can be charged.

- *Connect to the Controller*: Connect to the charge controller, ensuring the wiring is protected from the elements and secured against the tower to prevent damage. Install near the battery bank to minimize voltage drop.

- *Connect to the Battery Bank*: Wire the charge controller to your bank, following the manufacturer's guidelines. The charge controller regulates the power flow to the batteries, preventing overcharging. Connect the charge controller's output to the battery bank's terminals. Make sure to match the polarity (positive to positive, negative to negative) to avoid damage.

- *Inverter Connection*: From the battery bank, connect to the inverter, which will convert the DC power to AC power for use in your home.

Ensure electrical connections are tight at every step and use appropriate gauge wiring to handle the expected power load. Installing a disconnect switch between the turbine and the rest of your system is also wise for easy maintenance and safety. Ensure the voltage and current ratings are suitable for the maximum output of your wind turbine and the maximum load of your system. It's essential to consider both the voltage and current to ensure cables and connectors can handle the power without overheating or degrading.

Step 5: Test and Commission

Before you start enjoying your new wind power, a thorough testing phase ensures everything works as it should. Here are the steps you should follow:

- *Visual Inspection*: Check every connection, ensuring they're tight and correctly wired. Look for any signs of damage to panels, wires, or the inverter.

- *Voltage Testing*: Using a multimeter, verify that your turbine is generating the expected voltage. Check at various points—output, rectifier, inverter input, and inverter output—to ensure no unexpected drops occur.

- *Inverter Testing*: Most inverters have a testing mode you can initiate. This checks the inverter's ability to convert DC to AC efficiently and to shut down in case of a grid failure, a safety requirement known as anti-islanding.

- *System Monitoring Setup*: Finally, configure any system monitoring tools you have. These could be apps or online platforms that track your system's performance. They're invaluable for catching issues early and optimizing your setup.

This meticulous, step-by-step approach ensures your DIY wind turbine installation not only meets your energy needs but does so safely and efficiently.

Wrapping It Up...

As we conclude this chapter, it's clear that setting up a wind turbine is not just a technical challenge but also a step towards embracing renewable energy, reducing your environmental impact, and achieving a certain level of autonomy from the conventional power grid. With the foundation in place, the turbine towering against the sky, and the blades spinning in the wind, you've transformed the landscape in more ways than one.

Next, we'll focus on system maintenance to ensure that your system operates at optimal efficiency.

7

Maintain Peak Performance

Imagine your off-grid power setup as a living, breathing entity. Just like any living thing, it thrives with a bit of TLC. Without regular check-ups and maintenance, even the most robust solar or wind system can show signs of wear, much like an untended garden. This chapter guides you to cultivating a system that survives and flourishes, ensuring your energy supply remains as steady and reliable as the rising sun.

Solar

Dirt, dust, bird droppings, and leaves are not just eyesores; they're performance robbers, shading your panels and reducing their efficiency. Here's how to keep them spotless:

- *Frequency*: How often you need to clean your panels varies by location. Monthly cleanings might be necessary in dusty areas or places with frequent bird visits. Otherwise, a bi-annual check and cleaning as needed should suffice.

- *Methods*: A soft brush or cloth, a hose, and some mild soap are all you need. Avoid abrasive materials that can scratch the panels. A professional cleaning service might be the best call for those hard-to-reach places or if you're dealing with stubborn grime.

- *Safety First*: If your panels are on the roof, consider your safety. Use a sturdy ladder, wear nonslip shoes, and, if in doubt, leave it to the pros.

Inspect Mounting Systems

Mounting systems are the unsung heroes of your solar and wind power systems. They keep your panels or turbines in place, but they're not immune to the elements. Follow these steps to ensure your mounts are in perfect working order:

- *After Severe Weather*: Give them a once-over after heavy storms, snow, or high winds. Look for loose bolts, corrosion, or any signs of wear.

- *Rust and Corrosion*: Checking for these is especially important if living in a wet, humid, or coastal area where moisture can accelerate rust and salt can speed up corrosion. Rust-proofing can extend their lifespan.

- *Structural Integrity*: Ensure the mounts are straight and level. Any tilting could indicate issues with the foundation or the structure itself.

Monitor System Performance

Keeping an eye on how your system is doing isn't just about catching issues early; it's about understanding your energy habits and optimizing accordingly. Here are a few things to consider when monitoring your system's performance:

- *Monitoring Systems*: Many modern setups have monitoring software that provides real-time energy production and consumption data. Use it.

- *Interpreting Data*: Notice a sudden dip in energy output? It could be a sign something's amiss, from shading to equipment failure. Regularly compare your system's performance against previous months and seasons to spot any inconsistencies.

- *Manual Checks*: Don't rely solely on tech. A visual check, listening for unusual noises, or simply being aware of changes in your system's behavior can catch issues no software can.

- *Panel Degradation*: Note that the efficiency of solar panels will gradually decrease over time due to degradation. On average, a quality solar panel will degrade at 0.5-3% annually throughout its lifespan.

Battery Health Checks

Batteries are the heart of your storage system, and like any heart, they need care to keep the energy flowing smoothly. To ensure your batteries stay healthy, here are some things to watch out for:

- *Visual Issues*: Inspect your batteries and look for signs of bulging, leakage, or corrosion at the terminals. These can be early warnings of battery failure.

- *Voltage and Specific Gravity*: For those who like to get hands-on, checking the voltage and, for lead-acid batteries, the specific gravity of each cell can provide insights into their health. A multimeter and a hydrometer are your friends here.

- *Charge Levels*: Overcharging or letting batteries run too low can shorten lifespan. Ensure your charge controller is set correctly to keep battery charging within safe limits.

- *Temperature*: Batteries hate extremes. Too hot or too cold can affect their performance and longevity. If your battery bank is outdoors, consider insulation or a protective enclosure to moderate temperature fluctuations.

Maintenance isn't just about preventing breakdowns; it's about efficiency, longevity, and reliability. A well-maintained system performs better and can save you money in the long run by extending the life of your components. Plus, there's something deeply satisfying about knowing your off-grid system is running at its best, thanks to your efforts. So, grab that brush, that multimeter, or just your keen observer's eye, and give your system the care it deserves. It'll thank you for years of reliable, clean energy.

Troubleshoot Common Issues

When your solar system starts acting up, acting quickly can prevent a hiccup from becoming a headache. Let's dive into the nitty-gritty of identifying and fixing some of the most common issues that might dim your system's shine.

Identify Underperformance

Sometimes, your solar system might not be living up to its potential, and pinpointing the problem is the first step toward a solution. Here's how to spot and understand underperformance:

- *Check Your Energy Bills*: A noticeable increase in your energy bills could signal that your system isn't producing as much power as it should.

- *Compare Production Data*: If you have historical production data, compare current output to past months or the same month in previous years. Significant drops in production warrant a closer look.

- *Visual Inspection*: Sometimes, the issue is as simple as an obstruction blocking your panels. A quick visual inspection can reveal if fallen debris, snow, or unexpected shading is at fault.

Understanding these signs helps you catch issues early, saving you from more significant problems.

Inverter Problems

Inverters are crucial for converting solar energy into usable electricity, and when they falter, your whole system feels the impact. Here are some tips for tackling common inverter issues:

- *Error Codes*: Most inverters display error codes that can clue you into what's going wrong. Refer to your inverter's manual to decipher these codes and understand the next steps.

- *Power Output Issues*: If your inverter's power output doesn't match expectations, check for loose or damaged cables first. Sometimes, the solution is as simple as reconnecting a cable or replacing a worn-out wire.

- *Resetting Your Inverter*: Like many technological devices, an inverter sometimes needs a good reset. Turning it off, waiting a few minutes, and then turning it back on can often clear up minor glitches.

Keeping a keen eye on your inverter and understanding how to respond to its signals ensures it continues to do its job efficiently.

Battery Issues

Batteries store the sun's energy when you need it most. Still, they can run into sulfation, imbalance, and capacity loss. Here are some common issues and how to tackle them:

- *Sulfation*: This occurs when batteries are left undercharged for too long. Using a battery desulfator or ensuring your batteries are fully charged periodically can help prevent sulfation.

- *Imbalance*: Over time, batteries in a bank might start to charge and discharge unevenly. Regularly measuring and comparing the voltage of each battery can identify imbalances. As per the manufacturer's instructions, equalizing charges can help correct this.

- *Capacity Loss*: Batteries naturally lose their ability to hold a charge over time. Suppose your batteries are aging, and your system's storage capacity isn't what it used to be. In that case, it might be time to consider replacements.

Proper battery maintenance extends their life and ensures your storage system remains reliable.

Connectivity and Wiring Issues

Wiring is the circulatory system of your solar setup, and any issues here can lead to underperformance or even safety hazards. Here's what to look out for to ensure your system's wiring is in top shape:

- *Loose Connections*: Regularly check connections between panels, the inverter, and the battery bank. Tighten any loose connections with the appropriate tools.

- *Corroded Wires*: Inspect wires for signs of corrosion, especially in damp environments or near battery banks where acid corrosion can occur. Replace any corroded wires to prevent loss of efficiency and potential short circuits.

- *Damaged Insulation*: Look for signs of wear or damage to the insulation of your wires. Exposed wires can lead to short circuits and are a fire hazard. Repair or replace damaged wiring immediately.

By staying vigilant and addressing these common issues promptly, you ensure your solar system continues to operate smoothly, providing clean, renewable energy to power your off-grid life.

Wind

Wind turbines, those towering symbols of renewable energy, demand respect and care. Like any hard-working equipment, they perform best when appropriately maintained. This section will guide you through the essential upkeep practices, ensuring your wind turbine remains a steadfast clean energy provider.

Regular Inspections

Consistency is vital when it comes to inspecting your wind turbine. These giants face the brunt of nature's forces daily, and wear is inevitable. Scheduling monthly, biannual, and annual inspections can help catch issues before they escalate. What to focus on:

- *Blades*: Look for cracks, erosion, or any signs of wear. Blades are the workhorses of your turbine, and their condition directly impacts performance.

- *Tower*: Check the tower for rust, corrosion, or structural abnormalities. Remember, the tower holds everything up, and its integrity is paramount.

- *Mechanical Parts*: Gears, bearings, and bolts should be checked for wear and lubrication. These parts ensure the smooth operation of your turbine, translating wind into usable energy.

A meticulous inspection routine can prevent minor issues from becoming major problems, safeguarding your investment and energy supply.

Lubrication and Mechanical Maintenance

Moving parts and harsh conditions mean regular lubrication and mechanical maintenance aren't just recommended but necessary. Here's a schedule and some tips to keep everything running smoothly:

- *Lubrication Schedule*: Every six months, lubricate the turbine's moving parts. Use manufacturer-recommended lubricants to avoid any compatibility issues.

- *Gearbox Checks*: Annually inspect the gearbox, a critical component for energy conversion. Look for signs of wear or oil leakage and address them promptly.

- *Bearing Care*: Bearings reduce friction in your turbine, and their maintenance is crucial. Check for smooth operation and listen for any grinding noises that indicate wear.

Adhering to a maintenance schedule ensures each component functions optimally, extending the life of your turbine and maintaining its efficiency.

Storm and Weather Preparations

Wind turbines are built to withstand nature's moods, but extra precautions can minimize weather-related damage. Here's how to prepare for whatever the skies may bring:

- *Lightning Protection*: Ensure your turbine and tower have adequate lightning protection. This system diverts lightning strikes safely into the ground, protecting the electrical components.

- *Secure Loose Items*: Before a storm hits, block or remove any loose items from the vicinity of the turbine. High winds can turn these into projectiles, posing a risk to the turbine's structure.

- *Emergency Shutdown*: Know how and when to perform an emergency turbine shutdown. High wind conditions can sometimes exceed the turbine's design limits.

By taking these precautions, you can protect your turbine from severe weather events, ensuring it can continue to generate power once the storm passes.

Performance Monitoring

Staying informed about your wind turbine's performance allows for timely adjustments and maintenance, ensuring optimal operation. Here's how to keep a close eye on how well your turbine is doing:

- *Monitoring Tools*: Many turbines have monitoring software that provides real-time data on power output, wind speed, and turbine efficiency. Make the most of these tools.

- *Manual Observations*: Sometimes, there's no substitute

for seeing and hearing things yourself. Regularly observe your turbine in operation. Unusual sounds or vibrations can be early warning signs of issues.

- *Performance Benchmarks*: Set performance benchmarks based on historical data. This helps quickly identify deviations from expected output, signaling when it might be time for maintenance or troubleshooting.

By closely monitoring your turbine's performance, you can ensure it continues to operate efficiently, providing reliable, renewable energy to your off-grid system.

Maintaining a wind turbine involves regular inspections, scheduled maintenance, preparations for adverse weather, and diligent performance monitoring. This careful attention to upkeep maximizes the turbine's output. It extends its service life, ensuring it remains a valuable asset in your renewable energy arsenal.

Troubleshoot Common Issues

When your wind turbine starts acting up, it's like a finely tuned instrument hitting a sour note. It can be frustrating, but with some detective work, you can often get things humming again. Here's how to troubleshoot some common issues that might be causing trouble.

Power Output Fluctuations

Power output fluctuations in your wind turbine can stem from various sources, ranging from environmental factors to technical glitches. Pinpointing the root cause is the first step to smoothing out those energy highs and lows. Here's what to watch out for:

- *Wind Variability*: Check local wind speed data; natural fluctuations might be the culprit.

- *Blade Integrity*: Inspect the blades for any signs of damage or buildup that may be affecting performance. Even minor changes in blade integrity can impact efficiency.

- *Electrical Connectivity*: Examine the connections between the turbine and the battery bank. Loose or corroded connections can lead to inconsistent power output.

Familiarizing yourself with your setup's typical wind patterns and performance benchmarks can also provide valuable clues to whether the issue is systemic or an anomaly.

Blade Damage

Blades are the frontline warriors of your wind turbine, facing down everything from blistering winds to inadvertent bird strikes. Keeping them in top shape is vital for optimal performance. Here's how to spot and deal with blade damage:

- *Visual Inspection*: Regularly check your blades for nicks, cracks, or other damage. If your turbine is too high for a close inspection, use binoculars or a drone.

- *Minor Repairs*: Small cracks or holes can sometimes be fixed with epoxy resin or a specialized blade repair kit. Always follow the manufacturer's guidelines to ensure repairs are safe and effective.

- *Professional Help*: It's wise to call in the experts for more significant damage. Attempting major repairs without the proper skills or equipment can lead to further damage, and you may injure yourself.

Keeping blades well-maintained boosts efficiency and extends their lifespan, saving you time and money in the long run.

Vibration and Noise Problems

Unusual vibrations or noise from your wind turbine are more than just a nuisance; they can be harbingers of mechanical issues. If you notice your turbine is vibrating or making more noise than usual, here are a few things you can do:

- *Tighten Connections*: Start by ensuring that bolds and connections are screwed tight, especially those securing the turbine to the tower. Vibrations often stem from loose parts.

- *Balance Your Blades*: Imbalanced blades can cause vibrations, leading to noise and wear on the turbine. Balancing kits or professional services can correct this issue.

- *Lubrication*: Regularly lubricate the turbine's moving parts to minimize noise from friction. Consult your turbine's manual for the correct type of lubricant.

Promptly addressing these issues can prevent minor annoyances from escalating into significant disruptions.

Connectivity and Wiring Issues

Electrical issues can be particularly irritating, often hiding out of plain sight. Here's how to track them down:

- *Generator Check*: Ensure the generator is functioning correctly. A multimeter can help you verify its output.

- *Wiring Inspection*: Look for any signs of wear, corrosion, or damage to the wires running from your turbine to the battery bank. Replace any compromised wiring immediately.

- *Battery Connectivity*: Verify that all connections to the battery bank are secure and corrosion-free. Poor connections can lead to charging issues and reduced power storage.

Regularly reviewing your system's electrical components can prevent many common issues, ensuring a steady power flow from your turbine.

Wrapping It Up...

In wrapping up, remember that troubleshooting your systems isn't just about fixing what's broken; it's about understanding and optimizing your setup for long-term performance. From smoothing out power output fluctuations and repairing blade damage to reducing vibrations and addressing electrical issues, each step you take enhances your system's functionality and deepens your connection to the renewable energy journey.

BIOMASS

8

Alternative Energy Sources

Beyond conventional solar and wind solutions are various innovative renewable energy options for off-grid lifestyles. From the warmth of geothermal energy to the power of micro-hydro systems and the potential of biomass, we uncover the diverse possibilities for harnessing nature's resources.

Micro-Hydro

At its heart, a micro-hydro system involves channeling flowing water through a turbine to generate electricity. The components are straightforward: a water intake to direct flow, a penstock (a fancy term for a pipe) to carry water to the turbine, the turbine itself where the magic happens, a generator for converting mechanical energy into electricity, and the electronics to manage and distribute the generated power.

The beauty of micro-hydro lies in its efficiency. Unlike solar and wind, water flows around the clock, offering a steady energy source. Plus, it doesn't need vast volumes; even a small stream can power a home if you play your cards right. The trick is capturing the kinetic energy of moving water, directly related to the water flow (how much water is moving) and head (the height from which the water falls). Higher flow and greater head mean more energy.

Site Assessment

Finding the right spot for a micro-hydro system involves a bit of legwork, but it's well worth the effort. Here's what to look out for:

- *Stream Location*: Ideal spots are where the stream has a natural drop, or the land allows for a drop.

- *Flow Rate*: More water moving means more potential power. You can measure flow rate by timing how long it takes to fill a container of known volume.

- *Head*: This is the vertical drop available for your system. A simple way to measure this is by using stakes and a level over the distance of your intended penstock.

Tools like Google Earth can help with initial assessments, but getting your boots muddy by visiting the site is irreplaceable. Sometimes, the best spots are those you'd least expect.

Pros and Cons

Micro-hydro systems come with compelling advantages, but it's not all smooth sailing. Here are some points to consider:

Advantages:

- *Consistent Power Output*: Since water flows day and night, micro-hydro systems can provide a constant energy source, a significant advantage over solar and wind.

- *High Efficiency*: Water is dense and carries a lot of energy, making micro-hydro systems incredibly efficient.

- *Low Impact*: Once installed, micro-hydro systems have a minimal environmental footprint, especially compared to larger hydroelectric projects.

Disadvantages:

- *Initial Investment*: Setting up a micro-hydro system can be costly, mainly due to the civil engineering work required.

- *Permitting and Regulations*: Water rights and environmental regulations can be tricky to navigate, depending on your location.

- *Environmental Considerations*: Though generally low impact, diverting water can affect local ecosystems, which requires careful planning and possibly mitigation measures.

Case Studies

Real-life examples highlight the transformative potential of micro-hydro systems. Consider a family living in a remote area, their home powered by a stream that runs through their property. With a modest setup, they generate enough electricity to power their home, charge electric vehicles, and even return excess power to the grid, turning a natural resource into a boon for sustainable living.

Another case features a small community that collaborated to install a communal micro-hydro system. By pooling resources and sharing the generated electricity, they have created a self-sustaining microgrid that powers several homes, reduces their collective carbon footprint, and fosters a sense of community around shared sustainability goals.

These stories are more than just tales of ingenuity; they're blueprints for what's possible when we tap into the power of flowing water. With the right approach, a stream or river can become a cornerstone of off-grid energy independence, offering a reliable, renewable power source that runs day and night.

From the technical nitty-gritty of understanding how these systems work, assessing potential sites, and weighing the pros and cons to drawing inspiration from those who've successfully harnessed the power of flowing water, micro-hydro power embodies the essence of renewable energy: turning natural processes into partners in our quest for a sustainable, off-grid life.

Biomass Energy

Turning the remnants of yesterday's garden or the leftovers from a harvest into today's energy might sound like a magic trick. Yet, it's a reality made possible by biomass energy. This method of generating heat or electricity from organic material is as old as the first campfires. Still, it has evolved into a sophisticated component of modern off-grid living. Here, we explore how the oldest forms of energy known to humanity are being repurposed for the cutting-edge needs of off-grid life.

Biomass energy harnesses the chemical energy stored within organic material. When these materials, such as wood chips, plant residues, or even animal waste, are burned or decomposed, they release this stored energy as heat, which can be used directly or converted into electricity. It's a process that mirrors natural ecological cycles, where dead matter provides the energy needed for new life.

Types of Biomass Fuels

The variety of materials that can be used as biomass fuels is vast, and each has its own unique characteristics and uses. Let's look at three common types:

- *Wood*: The most traditional biomass fuel, wood, can be used in its raw form, as pellets, or as wood chips. It's ideal for heating and has a high energy content.

- *Agricultural Residues*: Leftover materials from farming, like straw or corn husks, can be burned for heat or processed into biofuels.

- *Biogas*: Created through the anaerobic digestion of organic matter by bacteria, biogas is a mix of methane and carbon dioxide. It can be used to generate electricity, cook, or heat water.

Each fuel type offers different benefits and challenges, from the high heat output of wood to the renewable cycle of biogas production. Which one you choose often depends on what's available and sustainable in your location.

Biomass Fuels

Off-Grid Homes

For those living off the grid, biomass energy can be a game-changer, providing heating, electricity, and even cooking gas through several technologies:

- *Wood Stoves*: Simple yet effective, modern wood stoves offer efficient heating with minimal pollution, making them a favorite in cold climates.

- *Biomass Boilers*: Similar to traditional boilers but designed to burn biomass fuels, these systems can heat entire homes and provide hot water.

- *Biogas Digesters*: These systems ferment organic waste in a sealed container, producing biogas that can be used for cooking, heating, or generating electricity.

Implementing these systems requires careful planning to meet your energy needs without overwhelming your space. For instance, a wood stove might be perfect for a cozy cabin, while a biomass boiler might benefit a more prominent home.

Sustainability and Efficiency

The green credentials of biomass energy hinge on two main factors: the sustainability of the fuel sources and the efficiency of the system used. Here are some guidelines to optimize both:

- *Source Locally*: Using biomass materials produced close to home reduces transportation emissions and supports local economies.

- *Use Waste Products*: Opting for waste materials, like agricultural residues or the by-products of wood processing, turns potential landfill fodder into valuable energy.

- *Maintain Your System*: Regular cleaning and maintenance of biomass systems ensure they burn cleanly and efficiently, maximizing energy output and minimizing emissions.

Moreover, the design of your biomass energy system plays a crucial role in its overall efficiency. Advanced combustion technologies in wood stoves and boilers can dramatically reduce pollutants and improve energy conversion rates. Similarly, modern biogas digesters are designed to maximize gas production while minimizing residue.

Incorporating biomass energy into off-grid living provides a reliable source of heat and power. It aligns with the principles of sustainability and self-sufficiency. By thoughtfully selecting and managing biomass fuels and choosing and maintaining the right technology, you can enjoy the benefits of this ancient energy source in a modern, environmentally friendly way.

Geothermal Energy

Digging into geothermal energy unveils an efficient method to keep our homes comfortable, irrespective of the sweltering summers or bone-chilling winters. This section peels back the layers of the earth to reveal how its steady, underground temperatures can be a gold mine for regulating our living spaces.

Geothermal Principles

A few feet beneath the surface, the earth maintains a nearly constant temperature year-round, regardless of the weather changes above. This principle lies at the heart of geothermal heating and cooling systems. By tapping into these stable underground temperatures, geothermal systems utilize a loop of pipes buried in the ground to transfer heat to or from your home. The system extracts heat from the earth during winter to warm the house. In contrast, it reverses the process in summer, removing heat from your home and transferring it underground to cool it down. This exchange is facilitated by a heat pump, which concentrates the earth's thermal energy and then circulates it through your home.

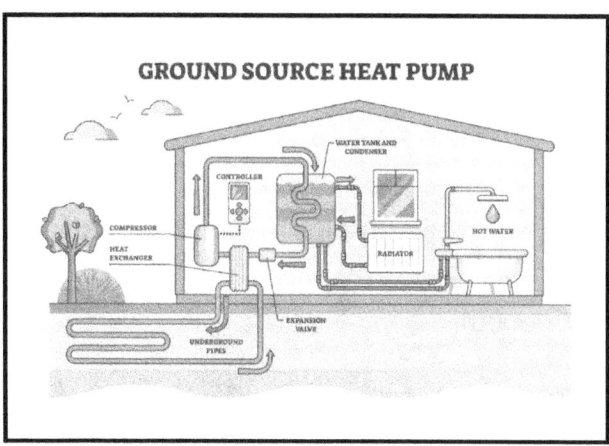

Geothermal Energy

Installation

The installation of a geothermal system is a task for professionals, involving a series of steps that ensure efficiency and longevity:

- *Initial Assessment*: A thorough evaluation of your property is crucial to determine the most suitable type of ground loop system (horizontal, vertical, or pond/lake) and its size. Factors such as soil condition, available space, and heating and cooling requirements are considered in this assessment.

- *Ground Loop Installation*: Depending on the assessment, trenches or boreholes are dug to lay the loop system. Horizontal loops are suitable for properties with ample yard space, while vertical loops are used where space is limited.

- *Heat Pump Installation*: The heat pump and distribution system (ductwork or radiant heating) are installed inside the home. This setup moves the heat to or from the ground loop and into your house.

- *Connection and Testing*: The ground loop is connected to the heat pump. The system is then filled with a fluid that facilitates heat transfer, and rigorous testing is conducted to ensure everything operates as it should.

The complexity of these steps underscores the importance of professional expertise in installing geothermal systems.

Costs and Benefits

Let's balance the books on geothermal energy by weighing its initial setup costs against the long-term financial benefits:

- *Upfront Costs*: The installation of geothermal systems can be pricey, often several times the cost of traditional HVAC systems. This includes the price of drilling, the ground loop, the heat pump, and any necessary modifications to your home's ductwork.

- *Long-Term Savings*: Despite the hefty initial investment, geothermal systems pay for themselves over time through significantly reduced heating and cooling bills. The efficiency of these systems means lower energy use, translating into savings of 30-60% on heating and 20-50% on cooling annually.

- *Incentives*: To sweeten the deal, various tax credits, rebates, and incentives to offset the installation costs of geothermal systems are available in many regions.

Environmental Impact

Choosing geothermal energy is a win for the planet as well. Here are some of its green credentials:

- *Reduced Greenhouse Gas Emissions*: By leveraging the stable temperatures of the earth, geothermal systems use less electricity than conventional HVAC systems. This reduction in energy consumption directly translates to lower greenhouse gas emissions.

- *Minimal Land Footprint*: Unlike wind or solar farms, geothermal systems have a minimal visual and physical footprint on the landscape. Most of the setup is underground, preserving your property's natural aesthetics and usability.

- *Sustainable Resource*: The heat from the earth is virtually inexhaustible, at least on human timescales. This makes geothermal energy one of the most sustainable options for heating and cooling, providing a consistent energy source without depleting the planet's resources.

Incorporating geothermal systems into our homes offers a practical application of the earth's natural warmth for our comfort. It is a testament to the innovative ways we can live harmoniously with our planet, tapping into its renewable resources to create sustainable and efficient homes. From the steady embrace of the earth's warmth to the whisper-quiet operation of a heat pump, geothermal energy offers a serene yet powerful solution to our heating and cooling needs, proving that sometimes, the best technologies work seamlessly with nature.

The Power of the Ocean

The ocean, covering more than 70% of our planet, is a colossal energy source, pulsing with potential that goes beyond the shores and deep into innovation. The rhythmic dance of the tides, the relentless push and pull of waves, and the vast thermal layers hidden beneath the surface offer a spectrum of opportunities to tap into clean, renewable energy. Here, we explore the intricate world of ocean energy technologies, shedding light on how the boundless forces of the sea can be transformed into a beacon of power for off-grid living.

Ocean Energy Overview

The ocean's energy comes in several forms, each with unique mechanisms for electricity generation. These include tidal energy, wave energy, and ocean thermal energy conversion (OTEC). Tidal energy leverages ocean tides' predictable rise and fall to generate power. In contrast, wave energy captures the surface motion of the sea. OTEC exploits the temperature difference between the warmer surface water and the colder deep sea water to produce electricity. Together, these technologies present a frontier for sustainable energy, promising to propel off-grid capabilities into new depths.

Tidal and Wave Energy

Tidal energy systems operate on the principle of capturing the kinetic movement of water during tide changes. This is typically achieved through underwater turbines that spin with the tide's flow, similar to wind turbines in air but beneath the water's surface. Strategic placement in locations with significant tidal ranges can optimize their effectiveness, offering a steady power output aligned with the natural ebb and flow of the ocean.

Wave energy, on the other hand, harnesses the up-and-down movement of sea waves. Devices like buoys, oscillating water columns, or attenuators convert this motion into mechanical energy, driving generators to produce electricity. The relentless nature of waves, especially in coastal regions, positions wave energy as a promising contender for off-grid energy solutions, capable of delivering power where it's most needed.

Ocean Thermal Energy Conversion Systems

OTEC stands out for its ability to generate electricity from the temperature difference between the ocean's warm surface water and its cold depths. This process involves circulating a working fluid with a low boiling point, such as ammonia, between heat exchangers. The warm surface water evaporates the fluid, driving a turbine connected to a generator. The vapor is then condensed using cold water from the ocean's depths, creating a continuous cycle of evaporation and condensation that generates electricity.

Ideal for tropical regions where the temperature difference is most pronounced, OTEC offers a glimpse into the future of ocean energy, providing a consistent and reliable energy source that could revolutionize power generation in off-grid communities.

Wrapping It Up...

In conclusion, exploring alternative off-grid energy sources such as micro-hydro, biomass fuels, geothermal energy, and ocean power unveils a promising horizon for sustainable and renewable energy. Each source offers unique advantages and represents a step away from fossil fuel dependency towards a greener, more sustainable energy future.

While solar and wind energy are popular off-grid options, alternative energy sources discussed in this chapter may well be viable depending on your location and circumstances.

9

Your Financial Roadmap

Transitioning to off-grid living involves more than just disconnecting from the grid. It requires careful financial planning to ensure that your move toward self-sufficiency is both sustainable and rewarding. This chapter serves as your compass and map, helping you navigate the financial challenges that come with going off-grid.

Test Your Knowledge

Before we start, here's a little quiz to test your knowledge of off-grid financial planning:

1. What is a key consideration when planning the installation of an off-grid system?

A. Choosing the color scheme for equipment

B. Predicting weather patterns for the next decade

C. Researching prices and budgeting for maintenance

D. Ensuring neighbors approve of the appearance

Correct Answer: C

2. What type of loan is specifically designed for eco-friendly projects, including off-grid systems?

A. Auto loans

B. Green loans

C. Personal loans

D. Mortgage loans

Correct Answer: B

3. What is one hidden cost of transitioning to off-grid living?

A. Permitting and legal fees

B. Subscription services

C. Landscaping modifications

D. Entertainment systems upgrades

Correct Answer: A

4. Which strategy is suggested for budgeting an off-grid transition?

A. Waiting for technology to advance

B. Starting with a small system and expanding as budget allows

C. Ignoring maintenance costs to save money

D. Buying all components at once for convenience

Correct Answer: B

5. How does leasing differ from buying off-grid system components?

A. Leasing requires a larger upfront investment

B. Leasing offers immediate access with little to no initial cost

C. Buying involves paying more over time

D. Buying limits system flexibility and upgrade options

Correct Answer: B

6. What is a benefit of utilizing financial planning tools for funding an off-grid system?

A. They guarantee approval for loans

B. They outline your budget and help evaluate loan options

C. They replace the need for a detailed itemized plan

D. They provide legal advice on permits

Correct Answer: B

7. Which of the following is true regarding government incentives for off-grid systems?

A. They exclusively offer non monetary support

B. They apply uniformly across all states

C. They can significantly lower initial and operational costs

D. They are only available to commercial properties

Correct Answer: C

8. What is a key factor to consider when deciding between DIY and professional installation of off-grid systems?

A. The color compatibility with existing structures

B. The potential for learning and personal empowerment

C. The availability of friends to help

D. The preference for outdoor vs. indoor work

Correct Answer: B

9. What factor significantly impacts the initial cost of an off-grid system?

A. The brand of equipment used

B. The size of the property

C. The type and number of batteries required

D. The distance from the nearest city

Correct Answer: C

10. How can crowdfunding and community support benefit an off-grid project?

A. By ensuring technical support from the government

B. By providing financial backing and building a supportive community

C. By automatically granting permits and legal approvals

D. By decreasing the importance of an itemized plan

Correct Answer: B

Scoring

- 10 Points: Nice job!

- 6-7 Points: Good basic knowledge

- 2-3 Points: Needs improvement

- 0 Points: Oh, dude, you really need this book!

Estimating Your Upfront Costs

When considering an off-grid power solution, the upfront cost is a significant factor. To better understand your potential financial commitment, let's examine estimated costs for solar and wind turbine systems tailored for off-grid homes.

Before we start, it's important to emphasize that the cost estimates shared here are just rough guidelines. They might change based on market conditions, your location, and the unique needs of your project. Plus, any potential tax incentives and rebates (which we'll cover later in this chapter), along with variable equipment costs, can really affect your total expenses. So, it's a good idea to do some thorough research and gather a few quotes before moving forward. Just a heads-up: these estimates are based on costs from January 2025.

For our estimates, we'll assume the following parameters:

Daily Energy Consumption 15 kWh

Power Output 3.5 kW

Lithium-ion Battery Bank

- Storage Capacity 10 kW

- Rated Power Output 3.5 kWh

Labor Installation = $4,000-$6,000

Solar Power Cost Estimate

Monocrystalline solar panels are adopted since they have the best efficiency.

Monocrystalline Solar Panels (3.5 kW system): $4,000–$6,000

Battery Storage (10 kWh, Lithium-Ion): $3,000-$4,000

Inverter (3.5 kW): $1,000-$1,500

Installation & Miscellaneous Costs: $4,000-$6,000

Total Estimated Cost: $12,000-$17,500

This solar system has sufficient battery capacity to supply power during the night. While boosting battery storage can enhance system reliability, it does come with added expenses. On the other hand, opting for a backup generator might be a more budget-friendly option. If you'd like more information about off-grid generators, check out Bonus Chapter 1: Generators!

Wind Turbine Cost Estimate

Wind Turbine (3.5 kW system): $3,500–$5,000

Battery Storage (10 kWh, Lithium-Ion): $3,000-$4,000

Inverter (3.5 kW): $1,000-$1,500

Installation & Miscellaneous Costs: $4,000-$6,000

Total Estimated Cost: $11,500-$16,500

The variation in cost estimates can be attributed to the prices of solar panels versus the wind turbine. Costs for other elements like batteries remain the same for both estimates. With recent advancements in renewable energy technology, we can look forward to even more cost savings and efficiency improvements for both solar and wind systems in the future.

Initial Costs vs. Long-Term Savings

Going off-grid is an investment in your future but requires upfront capital, like all investments. As we can see with our cost estimates, the initial costs can be substantial. Yet, these costs should be viewed through the lens of long-term savings. By freeing yourself from monthly utility bills, the systems pay for themselves over time, often in less than a decade. Moreover, considering the rising costs of traditional energy, the savings become even more significant as time passes.

By comparing these initial costs against potential savings on utility bills, the financial benefits of off-grid living begin to shine through. It's not just about the break-even point but about gaining energy independence and contributing to a more sustainable world.

Itemize Expenses

Using the estimated costs as a guide, you can create a more detailed itemized plan for your off-grid system. Here's a quick guide on how to craft your plan:

1. *List Every Component*: Include solar panels, wind turbines, inverters, batteries, wiring, and other essentials.

2. *Research Prices*: Look up current prices for each item, looking for deals or bulk purchase discounts.

3. *Add Installation Costs*: Whether you're going DIY or hiring

professionals, add estimated labor costs.

4. *Consider Maintenance*: Set aside a portion of your budget for regular maintenance and unexpected repairs.

This plan helps budget and secure financing, as it provides a clear picture of the investment needed.

Hidden Costs

Hidden costs are the boulders in the path of your off-grid journey. Identifying them early helps in navigating around financial surprises down the road. Some of these costs include:

- *Permitting and Legal Fees*: Depending on your location, you might need permits to install solar panels or wind turbines, which can add to your expenses.

- *Insurance*: Off-grid systems may affect your home insurance premiums. It's worth checking this out beforehand.

- *Transportation*: Delivering equipment can be costly if you live in a remote area.

Awareness of these potential hidden costs empowers you to budget more accurately and avoid being blindsided.

Budget Strategies

Finally, let's discuss budgeting strategies to make your off-grid dream a reality without breaking the bank. Here are some tips:

- *Start Small*: You don't have to go off-grid all at once. Begin with a small system and expand as your budget allows.

- *Phase Your Investment*: Plan your transition in phases, prioritizing components based on your energy needs and financial capacity.

- *DIY Where Possible*: Consider taking on some installation work yourself to save on labor costs, provided you have the skills and confidence to do so safely.

- *Prioritize Investments*: Focus on components that offer the best return on investment, such as energy-efficient appliances that reduce overall power needs.

By approaching your off-grid transition with a clear financial strategy, you'll confidently navigate the path, ensuring a smoother journey to energy independence and sustainable living.

As we've seen, the journey to off-grid living is much like charting a path through an unexplored forest. It requires preparation, a clear understanding of the terrain (or costs, in this case), and a solid plan to reach your destination. With the proper financial roadmap, the clearing beyond the trees—your sustainable, self-sufficient, off-grid home—is well within reach.

Finance Your Power System

When powering up your off-grid system, the financial fuel you choose can make all the difference. Not everyone has a treasure chest ready to pour into their off-grid setup, but worry not—there are several routes to secure the necessary funds, each with its own map and compass.

Loan Options

Navigating the sea of loan options can feel like finding a lighthouse in a storm. However, certain loans shine brighter for those looking to invest in off-grid systems. Green and home improvement loans stand out for their friendlier terms for eco-conscious projects. Here's a quick look at both options:

- *Green Loans*: Specifically designed for eco-friendly projects, these loans often come with lower interest rates and longer repayment periods. They're tailored for installing renewable energy systems, making them a perfect match for your solar panels or wind turbines.

- *Home Improvement Loans*: While not exclusively for green projects, these loans provide a broad umbrella under which off-grid systems fit comfortably. The key is demonstrating how your project boosts your property's value, a criterion off-grid improvements meet handily.

Each loan type opens different doors. Green loans whisper of long-term commitment to sustainability. In contrast, home improvement loans offer a more traditional path with a twist—investing in your home's energy independence.

Crowdfunding and Community Support

In a world where community often stretches beyond geographical boundaries, crowdfunding presents a novel way to gather the financial backing for your off-grid aspirations. Platforms like Kickstarter and GoFundMe allow you to share your vision with the world, tapping into a wellspring of potential support. Here's how you might navigate these waters:

- *Craft Your Story*: A compelling narrative that captures the essence of your off-grid project can resonate with like-minded souls, encouraging them to support your journey.

- *Offer Rewards and Updates*: Offering tangible rewards and regular updates keeps your backers engaged, turning them from mere contributors into a part of your off-grid adventure.

Crowdfunding pools financial resources and builds a community of supporters invested in your success. Additionally, pay attention to the power of local community support. Workshops, local environmental groups, or even a neighborhood gathering can spark interest and financial backing for your project.

Lease versus Buy

The crossroads between leasing and buying your off-grid system components is more than a financial decision; it's about weighing the scales of ownership against flexibility. Here's a glimpse into both paths:

- *Lease*: This option frees you from the upfront costs, handing you the keys to a functional off-grid system with little to no initial investment. The catch? You may pay more over time and have less control over the equipment, although the provider would be responsible for installation, repair, and maintenance. If leasing, you will not be eligible for state and federal rebates.

- *Buy*: Taking the plunge and purchasing your system outright demands a higher upfront investment but grants you full ownership, often resulting in lower costs in the long run. Plus, owning the system means you reap the full benefits of government incentives or rebates.

Both leasing and buying have their merits. Leasing offers an immediate step into off-grid living without the financial weight. At the same time, buying is a journey of building equity and independence.

Financial Planning Tools

With the myriad elements of financing an off-grid system, keeping track of everything without help can take time and effort. Thankfully, financial planning tools and software offer a lifeline. Whether it's budgeting apps, loan calculators, or energy savings estimators, these tools help you:

- *Outline Your Budget*: A clear view of your income, expenses, and savings goals makes allocating funds toward your off-grid project easier.

- *Evaluate Loan Options*: Loan calculators can demystify the long-term costs of different loans, helping you choose the one that aligns with your financial landscape.

- *Track Spending*: Real-time tracking of your spending against your budget keeps you on course, ensuring that your off-grid dream doesn't drift into financial fog.

From the first sketch of your off-grid blueprint to the final installation of a solar panel, these tools act as your financial compass, guiding you through the complexities of funding your off-grid system.

Maximize Government Incentives and Rebates

Living off the grid often starts as a dream fueled by a desire for independence and sustainability. However, transforming this dream into reality can be significantly aided by tapping into various governmental financial support systems designed to encourage renewable energy adoption. Across the board, from the corridors of federal buildings to local municipalities, there's a recognition that individuals like you contribute to a larger goal of environmental stewardship and energy independence. Let's explore how you can leverage these opportunities.

Federal and State Incentives

In the area of financial support, federal and state incentives are akin to sunlight for solar panels—essential and energizing. The government frequently offers tax credits and rebates, reducing the cost of installing renewable energy systems. For instance, the Federal Solar Investment Tax Credit (ITC) allows you to deduct a significant percentage of your solar system costs from your federal taxes.

State-level incentives can vary widely, with some states offering additional tax breaks, rebates, or even feed-in tariffs for surplus energy fed back into the grid. Researching what's available in your area is crucial, as these can substantially lower your initial and operational costs. Websites like the Database of State Incentives for Renewables & Efficiency (DSIRE) provide updated information on available incentives across the United States.

Grant Programs

Grants are like hidden treasure chests for off-grid projects, offering funding that can be paid. Several federal agencies, including the Department of Agriculture, offer grant programs supporting renewable energy projects, especially in rural and agricultural settings. These grants can cover various costs, from purchasing equipment to installation.

Local governments and nonprofit organizations also offer grants for community-based projects focusing on education, implementation, and renewable energy improvement. These grants can significantly offset the financial burden of transitioning to off-grid living for individual homeowners and small communities.

Application Processes

Securing these financial benefits often involves navigating a labyrinth of application processes, each with its rules, deadlines, and required documentation. Starting early and staying organized is critical. Here's a simple guide to keep you on track:

- *Documentation*: Gather all necessary documentation ahead of time, including quotes for systems, proof of residence, and any required permits.

- *Deadlines*: Mark application deadlines on your calendar to ensure you get all the funding opportunities.

- *Follow-Up*: After submitting your application, don't hesitate to follow up with the agency for updates or provide additional information if requested.

Patience and persistence are your allies here, ensuring you successfully navigate the bureaucratic process to secure the financial support you need.

Case Studies

Real-world examples can illuminate the path and inspire your journey. Consider the story of a small community in Vermont that utilized state incentives and a federal grant to establish a microgrid powered entirely by renewable energy. Not only did the project reduce the community's carbon footprint, but it also provided a model for sustainable living that attracted attention and additional funding for future green projects.

Another inspiring case is a family in Colorado who transformed their ranch into an off-grid haven using solar panels and wind turbines. By leveraging state rebates and a USDA grant, they could cover a significant portion of their installation costs. Their story highlights the impact of thorough research and strategic application for incentives, turning their sustainable living dream into an affordable reality.

In both cases, successfully utilizing government incentives and rebates was pivotal, demonstrating the practical benefits of these financial support systems. These stories serve as beacons, guiding you toward realizing your own off-grid aspirations without bearing the total economic weight alone.

DIY versus Professional Installation

Deciding between rolling up your sleeves to install your off-grid system or hiring a professional is more than a matter of pride—a significant financial decision that merits a careful look. Let's break down the costs and what they entail.

Cost Comparison

When you decide to install yourself, the immediate cost benefit is clear: you save on labor, which can be a considerable portion of the total project cost. However, the equation is more complex. Materials are just one part of the puzzle. The time investment, potential for errors, and the cost of correcting those errors can offset initial savings.

While professional installation comes with higher upfront costs, it also comes with expertise and efficiency. Professionals can often secure materials at a lower price due to industry relationships and bulk purchasing, a saving that can be passed on to you. Furthermore, the risk of costly mistakes is significantly lower, and should something go wrong, it's on the contractor to fix it, often under warranty.

Pros and Cons of DIY

Taking the DIY route has its charms and challenges:

- *Cost Savings*: The most apparent advantage is the potential to save money, especially on labor costs.
- *Learning and Empowerment*: Installing your own system can be an enriching learning experience, offering a deep understanding of your energy system's workings.
- *Flexibility*: You can work at your own pace and on your own schedule.

Yet, there are notable drawbacks:

- *Time Commitment*: DIY installations can be time-consuming, especially for beginners.

- *Risk of Mistakes*: Without professional experience, costly errors increase, potentially negating any initial savings.

- *Safety Concerns*: Incorrect installation can pose significant safety risks, from electrical hazards to structural failures.

When to Hire Professionals

Specific scenarios strongly tilt the balance toward professional installation. Here's when and why you should hire a pro:

- *Complex Systems*: Professionals ensure a safe and efficient setup for intricate systems or those requiring specialized knowledge, like hybrid systems combining solar and wind power.

- *Regulatory Compliance*: Navigating permits and regulations can be daunting. Professionals are versed in local codes, ensuring your system meets all legal requirements.

- *Safety*: Professional expertise is crucial for projects involving electrical work or structural modifications to avoid accidents and ensure long-term protection.

Negotiate With Contractors

If you choose a professional, securing the best deal while ensuring quality work is critical. Here are some strategies:

- *Get Multiple Quotes*: This gives you a baseline for negotiation and ensures you're getting a competitive rate.

- *Check References and Reviews*: Past work can indicate a contractor's quality and reliability.

- *Discuss Materials*: Sometimes, you can save money by purchasing materials yourself or opting for less expensive alternatives recommended by the contractor.

- *Understand the Scope*: Be clear about what's included in the price and what might be extra. This clarity can prevent unexpected costs down the line.

In summary, whether you choose the satisfaction of DIY or the assurance of professional installation, both paths can lead to a successful off-grid setup. The key lies in understanding the full scope of each option's costs, benefits, and risks. With this knowledge, you're well-equipped to make the best choice for your situation, balancing cost, quality, and safety to achieve your off-grid dreams.

Wrapping It Up...

As we wrap up this exploration of the financial aspects of transitioning to off-grid living, it's evident that careful planning, informed decision-making, and a clear understanding of the costs involved are crucial. Whether crafting a detailed budget, navigating the maze of financing options, or choosing between DIY and professional installation, each step is a building block toward achieving energy independence in a financially sustainable way.

10

NAVIGATE THE LEGAL LANDSCAPE

Navigating the intricate web of the legal and permit requirements for setting up an off-grid power supply can often feel like an uphill battle against a seemingly impenetrable bureaucracy. This process is fraught with complex regulations, meticulous inspections, and an endless stream of paperwork that can test the patience of even the most stoic individuals. Yet, this frustrating journey through red tape is a necessary evil. It ensures that your off-grid system not only aligns with safety standards and environmental regulations but also secures the legal legitimacy needed for operation.

Adhering to these legalities, despite their cumbersome nature, is crucial not just for the protection of the individual setting up the system but also for the community and the environment at large. It guarantees that the transition to off-grid living contributes positively to sustainability goals and does not inadvertently harm local ecosystems or contravene public safety norms.

Test Your Knowledge

Before we start, here's a little quiz to test your knowledge of off-grid legal landscape:

1. What can zoning laws affect in relation to setting up an off-grid system?

A. Color schemes of buildings

B. Types of plants that can be grown

C. Land use and structure sizes

D. The number of animals allowed on the property

Correct Answer: C

2. Why are building codes important for off-grid living?

A. They ensure aesthetic compatibility with the neighborhood

B. They help to maintain high property values

C. They ensure structures and systems are safe and reliable

D. They prevent people from choosing off-grid living

Correct Answer: C

3. What is true about obtaining variances for off-grid setups?

A. They are automatically granted upon request

B. They require a public hearing and a strong justification

C. They are irrelevant for off-grid living

D. They apply only to commercial properties

Correct Answer: B

4. What common pitfalls should be avoided when applying for permits?

A. Applying for too many permits at once

B. Providing all the required information and documentation

C. Failing to provide all the required information or documentation

D. Taking too long to submit the application

Correct Answer: C

5. What factor can influence the cost of insurance premiums for off-grid homes?

A. The color of the home

B. The location and complexity of the off-grid system

C. The number of pets on the property

D. The age of the homeowners

Correct Answer: B

6. What is a key reason zoning laws are in place for areas considering off-grid living?

A. To limit the number of residents in an area

B. To enforce uniform architectural styles

C. To ensure land use is optimized for community benefit

D. To discourage off-grid living

Correct Answer: C

7. Which of the following permits is NOT typically required for an off-grid setup?

A. Noise exemption permit

B. Electrical permit

C. Building permit

D. Environmental permit

Correct Answer: A

8. What is an essential part of the permit application process for off-grid systems?

A. Relying on generic templates for all applications

B. Consulting with local authorities for specific requirements

C. Submitting applications without site surveys

D. Assuming approval is automatic

Correct Answer: B

9. When conducting a risk assessment for off-grid living, what is a crucial step?

A. Determining the entertainment options available

B. Identifying potential hazards and evaluating vulnerabilities

C. Assessing the visual impact of the off-grid setup

D. Calculating the daily energy consumption rates

Correct Answer: B

10. What factor can affect negotiations with insurance companies for off-grid homes?

A. The popularity of off-grid living in your area

B. The presence of a vegetable garden

C. Implementation of safety features and certifications

D. The number of previous claims made

Correct Answer: C

Scoring

- 10 Points: Nice job!
- 6-7 Points: Good basic knowledge
- 2-3 Points: Needs improvement
- 0 Points: Oh, dude, you really need this book!

Zoning Restrictions

Zoning laws, those invisible lines that dictate what can and cannot be done in specific areas, are the first hurdle to clear. They're there for a reason, often to ensure that land use is optimized for safety, health, and the overall benefit of the community. Here's what you might run into:

- *Land Use*: Many zoning laws have strict definitions of what land can be used for—residential, agricultural, commercial, etc. If your off-grid plans include specific agriculture or business types, ensure they align with local zoning.

- *Structure Size and Type*: There might be restrictions on the size or type of structures you can build. This can affect everything from the main house to outbuildings like greenhouses or workshops.

- *Utility Requirements*: Some areas require homes to be connected to the grid or have access to public sewer systems. Off-grid plans might clash with these requirements.

Navigating these restrictions starts with a visit to your local zoning office. They have the maps, the rules, and the advice you need to start on the right foot. It's also where you'll learn about applying for variances or exceptions if your off-grid dream needs to fit neatly into existing categories.

Building Codes

Building codes are all about safety. They ensure that structures are built to withstand environmental stresses, electrical systems are installed correctly, and plumbing doesn't pose a health risk. Here's a quick rundown:

- *Electrical Systems*: Off-grid doesn't mean unsafe. Electrical installations, solar panels, and battery banks must meet safety standards to prevent fire hazards or electrocution.

- *Structural Components*: Whether it's wind loading for turbines or the weight of snow on a solar panel array, making sure structures can handle environmental pressures is crucial.

- *Water and Sewage*: Proper disposal of waste and safe water supply systems are not just about comfort but preventing contamination and disease.

The key here is to work with certified professionals who understand off-grid systems and the local codes that apply to them. They can ensure your setup is both compliant and safe.

Engage With Local Authorities

Talking to local authorities about your off-grid plans can be daunting, but it's a step you must take. Here's how to make it a positive experience:

- *Schedule a Meeting*: Don't just drop in. Make an appointment with someone knowledgeable about zoning and building codes in your area.

- *Come Prepared*: Bring your plans, questions, and preliminary designs or ideas. Being prepared shows you're serious and respectful of their time.

- *Stay Positive*: Approach the meeting with a positive attitude. Remember, these folks are here to help, and they can be valuable allies in your off-grid journey.

Building a good relationship with local authorities can smooth the way for your project. They can offer insights, suggest alternatives, and even help you navigate the variance process.

Variances and Exceptions

Sometimes, the only way forward is to ask for an exception to the rules. Applying for a variance can be a complex process, but here's the gist:

- *The Application*: It usually involves filling out forms, paying fees, and providing detailed plans of your proposed project.

- *The Justification*: You must explain why your project should be granted an exception. This is where a well-thought-out presentation, showing how your off-grid home will be safe, environmentally friendly, and an asset to the community, comes into play.

- *The Hearing*: Many variance applications require a public hearing. This is your chance to present your case to the zoning board and the community.

- *The Decision*: After the hearing, the board will make a decision. If it's a 'no,' all is not lost. You can appeal, adjust your plans, or try a different approach.

Getting a variance is not guaranteed, but with a strong case and community support, it's definitely possible. Remember, zoning laws and building codes aren't there to stifle your off-grid dreams but to ensure that those dreams don't turn into nightmares for you or your neighbors. Compliance might seem like a hurdle, but it's just part of ensuring that your off-grid home is safe, legal, and sustainable for years.

Right Permits

Securing permits for an off-grid setup is like piecing together a puzzle where each piece represents a different requirement or regulation. It's about knowing which pieces you need and how they fit together to create a picture that local authorities approve of. Let's walk through the steps and insights necessary for this part of your project.

Permit Requirements

The kinds of permits you might need for your off-grid systems vary widely based on location and the specifics of your project. Generally, you're looking at:

- *Electrical Permits*: For installing solar panels, wind turbines, or any component that involves wiring and electrical work.

- *Building Permits*: Required for any new structures or significant modifications to existing ones, including installing specific off-grid systems.

- *Environmental Permits*: Depending on your system and its potential environmental impact, you may need permits relating to water use, waste management, or land disturbance.

Identifying the necessary permits is your first step, achieved by consulting with local building departments, zoning boards, and environmental agencies.

Application Process

Navigating the permit application process can be straightforward if approached methodically. Here is a step-by-step guide to keep you on track:

1. *Gather Information*: Before you apply, collect all necessary project details, such as plans, site surveys, and system specifications. This information will be crucial for completing your application accurately.

2. *Consult Authorities*: As mentioned, a preliminary chat with local authorities can clarify which permits you need and the criteria you must meet. This can save you time and effort in the long run.

3. *Submit Your Application*: With your information in hand and a clear understanding of requirements, fill out the application forms thoroughly. Include all requested details and attach any required documentation.

4. *Prepare for Inspections*: Many permits require an inspection of the proposed site or system. Ensure you're ready by reviewing what inspectors will look for and making any necessary adjustments beforehand.

5. *Stay Patient and Proactive*: After submission, the waiting game begins. However, staying proactive by checking in on your application status and being ready to provide additional information can help expedite the process.

Costs and Timelines

Understanding the financial and time commitments of obtaining permits is crucial for effectively planning your project. Here's what to anticipate:

- *Fees*: Permit fees can vary significantly based on your location and project scope. It's common for costs to range from a few hundred to several thousand dollars, so factor this into your budget.

- *Timelines*: The time it takes to approve your permits can also vary. Some might come through in weeks, while others take months. Early inquiries into expected timelines can help you align your project planning.

Planning financially and time-wise for these variables ensures you won't be caught off guard as your project progresses.

Common Pitfalls

Even with meticulous planning, it's possible to encounter hiccups in permitting. Being aware of these common pitfalls can help you steer clear of problems down the line:

- *Incomplete Applications*: Failing to provide all the required information or documentation is a surefire way to delay or deny. Double-check your application before submitting it.

- *Ignoring Local Regulations*: Every region has its unique set of rules. Not tailoring your project to fit these can result in permit rejections. Make sure your plans comply with local standards.

- *Underestimating Inspection Standards*: Inspectors follow strict criteria to ensure safety and compliance. Refraining from underestimating inspections' thoroughness can lead to failed inspections and delays.

By considering these pitfalls and planning accordingly, you can navigate the permit process more smoothly, keeping your off-grid project on track and within the bounds of local laws and regulations.

Insurance and Liability

Securing your off-grid haven requires more than just solar panels and rainwater catchment systems; it demands a solid safety net in insurance. This section returns the layers on ensuring your off-grid dream remains protected from unforeseen events. It's about peace of mind, knowing that should the unexpected occur, you're covered.

Insurance Coverage for Off-Grid Homes

Navigating insurance for an off-grid home is more complex than one might hope. Traditional policies might only partially encompass the unique aspects of off-grid living, leaving gaps in coverage. Here's a rundown of the types of insurance you should consider:

- *Property Insurance*: Shields your home and belongings from damage or loss due to fires, storms, or theft. However, the nuances of off-grid structures might require explicit mentions or additional riders in your policy.

- *Liability Insurance*: Living off the grid doesn't isolate you from legal liability if someone is injured on your property. Liability insurance is crucial for covering legal fees or medical expenses arising from such incidents.

- *Equipment Insurance*: Off-grid setups rely heavily on equipment that traditional home policies might not cover automatically. From solar panels to wind turbines and battery banks, ensuring these integral pieces are explicitly protected is vital.

Risk Assessment

Understanding the risks unique to off-grid living is the first step in mitigating them. Conducting a thorough risk assessment involves:

- *Identifying Potential Hazards*: From natural disasters pertinent to your location to the risks associated with your energy systems, pinpointing potential issues is critical.

- *Evaluating Vulnerabilities*: Assess how susceptible your off-grid home is to these hazards. For example, how might extreme weather affect your solar setup?

- *Prioritizing Risks*: Not all risks carry the same weight. Determining which poses the most significant threat helps focus your mitigation efforts and insurance coverage.

This proactive approach not only aids in securing appropriate insurance but also implements measures to minimize risks, such as installing lightning protection for solar arrays or reinforcing structures against storms.

Insurance Premium Factors

Several variables can sway the cost of insuring your off-grid home, sometimes making premiums higher than those of conventional homes. Here are vital factors insurers might consider:

- *Location*: Remote or disaster-prone areas can see higher premiums due to increased risk.

- *System Complexity*: The more complex your off-grid set-up, the higher the perceived risk, influencing premium costs. Simple, well-documented systems fare better.

- *Safety Measures*: Implementing safety features, such as firebreaks or emergency shut-offs for energy systems, can positively impact your premiums by reducing risk.

Understanding these factors lets you make informed decisions about your off-grid setup and negotiate with insurers.

Negotiate With Insurance Companies

Approaching insurance companies for off-grid coverage requires a blend of preparation and negotiation skills. Here's how to approach the task:

- *Research and Compare*: Start with a broad search of companies and policies. Some insurers might be more experienced with off-grid properties, offering better terms.

- *Highlight Safety Features*: When discussing your off-grid home, emphasize your systems' safety measures or certifications. This demonstrates a lower risk profile.

- *Leverage Professional Installations*: If your systems were installed by certified professionals, use this as a bargaining chip. It reassures insurers of the reliability and safety of your setup.

- *Ask About Discounts*: Ask about potential discounts for things like bundling policies or installing additional safety equipment.

The goal is to secure comprehensive coverage that acknowledges the unique aspects of your off-grid lifestyle at a fair rate. Remember, insurance is not just a mandatory hurdle but a critical component of your off-grid planning, ensuring that your sustainable dream home remains a source of joy and security for years.

Off-Grid Living and Legal Compliance: A Case Study

In the rolling hills of a rural county, where the maples turn fiery in autumn, the Johnson family set their sights on a life untethered from the grid. Their dream was simple yet bold: transforming a patch of land into a self-sufficient homestead powered by the sun and wind. However, their path to autonomy was dotted with legal hurdles, a testament to the intricate dance between innovation and regulation.

A Detailed Compliance Journey

The Johnsons' quest began with enthusiasm, tempered by the realization of the myriad regulations surrounding off-grid living. Their initial step was a deep dive into local zoning laws, which revealed a surprising restriction on stand-alone solar installations. Undeterred, they reached out to local officials, laying the groundwork for productive discussions.

Challenges arose when their building permit application hit a snag over concerns about their water purification system. The solution came as a detailed presentation to the zoning board, with expert testimonials and case studies demonstrating the system's efficacy and safety.

The Johnsons secured the necessary permits through perseverance and a willingness to engage. They fostered a dialogue about sustainable living within their community, paving the way for future off-gridders.

Lessons Learned

The Johnsons' journey underscored several vital lessons:

- *Early Research Pays Off*: Diving into regulations early on can uncover potential roadblocks, allowing for timely solutions or adjustments.

- *Community Engagement Is Crucial*: Open communication with local officials and neighbors can turn potential adversaries into allies.

- *Flexibility Is Key*: Being open to modifying plans or exploring alternative solutions can help navigate regulatory challenges more smoothly.

Their experience serves as a blueprint for others, emphasizing the importance of preparation, communication, and adaptability.

Success Factors

Several factors contributed to the Johnsons' success in navigating legal compliance:

- *Thorough Documentation*: Meticulous records and detailed plans bolstered their case, demonstrating a commitment to safety and environmental stewardship.

- *Professional Consultations*: Engaging with experts, from legal advisors to environmental engineers, gave the Johnsons the knowledge to address regulatory concerns convincingly.

- *Adherence to Safety Standards*: By prioritizing safety in the design and installation of their off-grid systems, they aligned their project with the core intent of building codes and zoning laws.

These elements not only facilitated their compliance journey but also enhanced the sustainability and efficiency of their homestead.

Future Considerations

Looking ahead, the Johnsons remain vigilant about potential regulation changes that could impact their off-grid lifestyle. They stay informed through local environmental groups and government bulletins, ready to adapt. Their proactive stance is a model for off-gridders everywhere, highlighting the importance of staying engaged with the evolving legal landscape of sustainable living.

Their story is a beacon for those contemplating a leap into off-grid living, illuminating the path through legal compliance with patience, engagement, and a steadfast commitment to safety and sustainability.

Wrapping It Up...

As we wrap up this exploration, it's clear that navigating the legal aspects of off-grid living is more than a checkbox on a to-do list. It's a critical step in ensuring your sustainable haven stands on solid ground, literally and figuratively. From the Johnsons' tale, we gather strategies for legal compliance and a deeper appreciation for the balance between individual initiative and community standards. Though filled with paperwork and permits, this journey is part of a larger story about transforming our relationship with the environment and each other. It's about building not just homes but futures as sustainable as free.

As we move forward, keep these lessons in mind. They're not just about navigating bureaucracy but about weaving the fabric of a lifestyle that respects the letter and spirit of the law, ensuring that our off-grid dreams contribute positively to the world we all share.

Bonus Chapter 1: Generators

Generators play a crucial role in off-grid living, serving as a reliable backup or supplemental power source when renewable energy sources are insufficient. This chapter will guide you through selecting the right generator for your off-grid system, understanding its operation, and maximizing its efficiency and lifespan.

Types of Generators

In the world of generators, variety abounds. Each type comes with its own set of advantages and disadvantages, tailored to different off-grid scenarios:

- *Diesel*: Known for their durability and high energy output, diesel generators are a popular choice for off-grid systems requiring significant power. They tend to have a longer lifespan and are more fuel-efficient than their gasoline counterparts. However, they can be more expensive upfront and louder during operation.

- *Gasoline*: Gasoline generators are widely available and typically have a lower initial cost than diesel or propane generators. They're suitable for smaller energy needs but can be less fuel-efficient and have a shorter lifespan. Gasoline also has a shorter shelf life, which can be a drawback for long-term storage.

- *Propane*: Propane generators run cleaner than diesel and gasoline models, producing fewer emissions. They're quieter and can have a longer shelf life when it comes to fuel storage. However, they may be less efficient in terms of energy output per gallon of fuel compared to diesel generators.

- *Dual-Fuel*: Offering flexibility, dual-fuel generators can operate on gasoline or propane, allowing you to switch between fuel types based on availability or preference. This versatility can be particularly advantageous in off-grid living, where fuel availability may vary.

- *Solar*: Solar generators are essentially large batteries charged by solar panels. They're quiet, emission-free, and ideal for smaller power needs. While not suitable as the sole power source for most off-grid homes, they're excellent for supplementing renewable energy systems and powering essential electronics.

Each generator type serves different needs, from the robust output of diesel to the clean, renewable energy of solar-powered options. The choice hinges on balancing factors like environmental impact, fuel availability, and power needs.

When choosing a generator, consider your energy needs, fuel availability, environmental impact preferences, and budget. First and foremost, calculate your peak energy usage and choose a generator that can comfortably handle this load. Next, consider what fuel types are readily available in your area. Sourcing locally is not only easier but can be cheaper. Then, weigh the initial cost against the long-term operation and maintenance costs and select a generator within your budget. Remember, a higher upfront cost could save you money down the line.

Size Your Generator

Getting the size right is crucial. If your generator is too small, it won't meet your energy needs; if it's too large, it will waste your fuel and money. Here's how to find the sweet spot:

- Calculate your peak load, which is the maximum power you'll use at any one time. Include essential appliances and occasional high-power needs.

- Consider your average consumption to understand your typical daily use.

- Aim for a generator capacity that comfortably covers your peak load while providing some headroom for unexpected needs.

This tailored approach ensures your generator is neither overtaxed nor underutilized, optimizing efficiency and longevity.

Fuel Efficiency and Storage

Fuel considerations are twofold: efficiency and storage. Efficient fuel use is paramount for cost control and minimizing environmental impact. Here's what to keep in mind when choosing a fuel:

- *Opt for Efficiency*: Diesel generators typically offer better fuel efficiency than propane, but technological advancements are closing the gap. Solar-powered generators shine here, with sunlight as their "fuel" source.

- *Plan for Storage*: Diesel and propane require safe storage solutions. For diesel, consider a well-ventilated, cool, dry place. Propane tanks should be stored outdoors, upright, and away from any potential sources of ignition.

- *Load Management*: Use energy-intensive appliances sequentially rather than simultaneously to avoid overloading the generator. This practice can extend the life of your generator and reduce fuel consumption.

Generators and Renewable Energy Systems

For a seamless off-grid experience, integrate your generator with your renewable energy system using a hybrid inverter or a transfer switch. This setup allows the generator to kick in automatically when renewable energy sources are insufficient, ensuring a continuous power supply without manual intervention.

Maintenance and Safety

Regular maintenance and a keen eye on safety can prevent accidents and extend your generator's life. Here are essential tips:

- *Regular Checks*: Routine inspections can catch issues before they escalate. Check fuel lines, connections, and batteries for signs of wear or damage.

- *Clean and Replace*: To ensure smooth operation, keep air filters clean and replace oil and spark plugs as recommended by the manufacturer.

- *Safety First*: Always operate generators in well-ventilated areas to prevent carbon monoxide buildup. Adhere to all manufacturer guidelines for operation and storage.

Following these guidelines ensures that your off-grid power system remains a reliable, efficient, and safe cornerstone of your sustainable living setup.

Wrapping It Up...

Selecting and utilizing a generator in an off-grid system is a balancing act between your energy needs, budget, and environmental considerations. By understanding the different types of generators available and following best practices for operation and maintenance, you can ensure a reliable and efficient power source that complements your renewable energy system, enhancing your off-grid living experience.

Bonus Chapter 2: The Complete Off-Grid Home

Living off-grid is becoming more feasible for those seeking a sustainable lifestyle. This chapter covers water systems, waste management, heating and cooling, and the role of technology in enhancing off-grid living. It envisions a world where every home is sustainable, innovative, and in harmony with the natural world.

Water Systems

Imagine turning on a tap in your off-grid home and pouring crystal-clear water, a direct gift from the skies above. Here, the age-old practice of collecting rainwater meets modern filtration techniques, ensuring that every drop is safe and precious. This chapter delves into making the most of every rainfall, purifying it to perfection, and using water in ways that give back to the earth as much as it sustains us.

Harvest Rainwater

Rainwater harvesting systems are a game-changer in off-grid living. They're about catching rain where it falls, storing it, and using it when needed. Here's how it works:

- *Collection*: Harvesting starts with your roof acting as a catchment area. Gutters funnel the rainwater into downspouts, leading directly to a storage tank. The larger your roof and the more efficient your gutters, the more water you can collect.

- *Storage*: Storage tanks can be above or below ground, depending on your space and needs. Dark, food-grade polyethylene tanks are famous for their durability and algae-prevention properties.

- *Filtration*: Before using the stored rainwater, it's vital to filter out debris, organic matter, and contaminants. Flush diverters and mesh filters at the point of the collection can help keep the bulk of unwanted material out of your storage tanks.

For anyone starting out, remember that the efficiency of your rainwater harvesting system hinges on regular maintenance. Cleaning gutters, inspecting tanks, and replacing filters are all in a day's work to ensure a safe water supply.

Water Purification Methods

Safe drinking water is non negotiable. Off-grid doesn't mean off-quality when it comes to water purification. Here are some methods suited for off-grid living:

- *Solar Distillation*: This purification method uses the sun's heat to evaporate water, leaving contaminants behind, before condensing it back into liquid form. It's practical and energy-efficient but requires sunny weather and patience.

- *Filtration Systems*: From ceramic filters to advanced reverse osmosis systems, various options suit different needs and budgets. Gravity-fed filters, which require no electricity, are particularly well-suited for off-grid setups.

- *Chemical Treatment*: Chlorine or iodine tablets can purify water, killing bacteria and viruses. It's a low-cost, reliable method, though some are wary of the taste and potential health impacts of long-term chemical use.

Choosing the proper purification method depends on your water source, the contaminants present, and your personal preferences. Often, a combination of methods ensures the safest drinking water.

Sustainable Water Practices

Living off-grid is an opportunity to rethink our relationship with water. Here are a few practices that can make a big difference:

- *Greywater Recycling*: Water from showers, sinks, and laundry doesn't have to go to waste. With proper filtration, it can be reused to water gardens or flush toilets.

- *Efficient Irrigation*: Drip irrigation systems deliver water directly to the roots of plants, minimizing waste and maximizing growth. It's an intelligent way to keep your garden thriving with minimal water.

- *Rain Gardens*: These are designed to capture runoff water and allow it to soak into the ground. They not only conserve water but also support local flora and fauna.

Every drop counts, and with these practices, you're not just saving water; you're actively contributing to the health of the ecosystem around you.

Regulatory Considerations

Before diving headfirst into collecting and purifying water, there's one more stream to navigate–regulations. Water laws vary widely, so it's crucial to:

- *Check Local Regulations*: Some areas restrict rainwater harvesting or require permits for specific systems.

- *Health and Safety Standards*: If you're purifying drinking water, ensure your methods meet local health and safety standards to avoid risks.

Staying informed and compliant keeps you on the right side of the law and ensures your water system is safe and sustainable.

We've explored how rainwater harvesting systems transform precipitation into a reliable water source, delved into the methods for purifying water to make it safe to drink, and highlighted sustainable practices that make every drop count. From the basics of collecting rainwater to the intricacies of legal compliance, it's clear that managing water off the grid is an exercise in responsibility, innovation, and respect for nature's lifelines.

If you are seeking independence from the grid but unsure how to secure a sustainable water supply, then another book in the *Off The Grid Survival* Series will give you the answers: *Off The Grid Prepper's Water Survival Plan*.

Waste Management Solutions

Living off the grid inspires a closer relationship with our environment, urging us to reconsider how we consume resources and dispose of or repurpose our waste. In this light, effective waste management becomes not just a necessity but a commitment to sustainability and respect for nature.

Composting Toilets

One of the most ingenious solutions to waste management in off-grid settings is the composting toilet. Far from the rudimentary outhouses of old, modern composting toilets offer a sophisticated, odor-free system that turns human waste into compost, a nutrient-rich material beneficial for the soil.

- *How They Work*: Composting toilets use the natural process of decomposition and evaporation to break down waste. Aerobic bacteria digest the waste, transforming it into compost. Ventilation systems ensure the process is odorless and the end product is safe for the environment.

- *Benefits*: The perks of composting toilets extend beyond their environmental friendliness. They're waterless, crucial in off-grid living, where water conservation is critical. Additionally, the resulting compost can enrich the soil on your property, closing the loop on waste.

Biogas Digesters

Biogas digesters present another innovative approach to managing organic waste, from kitchen scraps to animal manure. These systems produce methane, a clean-burning gas that can fuel stoves, heaters, or even generators.

- *How They Work*: In a sealed container, anaerobic bacteria digest organic waste, producing methane and carbon dioxide. This biogas is then captured and stored for use as fuel.

- *Benefits*: Beyond waste reduction, biogas digesters provide off-grid homes with a renewable energy source. They're a testament to the idea that waste need not be wasted but can be a valuable resource.

Recycle and Repurpose

Adopting a mindset of recycling and repurposing helps minimize the waste that off-grid homes contribute to landfills. It's about seeing potential in what's discarded and finding new uses for old items. Here's how to achieve this:

- *Creative Repurposing*: Items that have outlived their original use can be transformed through repurposing. Pallets become furniture, glass jars turn lanterns, and scrap metal morphs into art. The possibilities are as limitless as one's creativity.

- *Recycling*: Proper recycling ensures that materials that cannot be repurposed are processed and made into new products. Setting up separate bins for paper, plastics, metals, and glass helps streamline this process.

Waste Disposal Regulations

Navigating the waste disposal regulations is crucial for maintaining harmony with the environment and your community. Ensuring compliance protects nature and upholds the health and well-being of those living off the grid and their neighbors. To make sure you're in line with regulations, do the following:

- *Stay Informed*: Local regulations can dictate everything

from the legality of composting toilets to the requirements for waste disposal. Familiarizing yourself with these rules is the first step toward compliance.

- *Best Practices*: Even without strict regulations, adhering to best practices in waste management—such as proper composting techniques and safe biogas production—ensures your efforts are practical and environmentally responsible.

In off-grid living, managing waste is as much about embracing sustainability as innovation and resourcefulness. These solutions reflect a profound respect for the natural world, from turning waste into compost that feeds the earth to capturing biogas that powers our homes. They remind us that living off the grid is not merely disconnecting from municipal services but forging a deeper, more sustainable connection with the environment.

Smart Off-Grid Homes

In a world where technology touches almost every aspect of our lives, it's no surprise that even off-grid homes are getting smarter. The essence of off-grid living is not just about disconnecting from municipal utilities but crafting a lifestyle that's efficient, sustainable, and harmonious with nature. This is where intelligent technologies come into play, blending the rustic allure of off-grid living with the convenience and efficiency of modern automation.

Energy Management Systems

The heart of any smart off-grid home is its energy management system. This intelligent nexus acts as the brain for your power supply, seamlessly juggling inputs from solar panels, wind turbines, or other renewable sources. It's about ensuring that every watt of energy produced is used to its fullest potential, whether charging battery banks during times of surplus or diverting power to essential appliances when generation is low. These systems can:

- Predict energy needs based on usage patterns.

- Optimize battery life by preventing overcharging or excessive discharging.

- Automatically switch between energy sources to maintain efficiency and reliability.

The beauty of these systems lies in their ability to learn and adapt, ensuring that your off-grid home remains powered, even under changing conditions.

Automated Water Systems

Water, the lifeblood of any home, takes on even greater significance in an off-grid setting. Automated water systems revolutionize collecting, purifying, and using this precious resource. From intelligent rainwater harvesting systems that activate collection based on weather forecasts to purification setups that adjust filtration rates based on water quality sensors, automation ensures that every drop is used judiciously. Some notable advancements include:

- Intelligent irrigation systems that water your garden precisely when needed, reducing waste.

- Automated greywater recycling systems that filter and redirect water for non potable uses, lessening the demand on your main water supply.

These systems not only conserve water but also reduce the manual effort involved in managing your off-grid water supply, allowing for more time to enjoy the tranquility of your surroundings.

Smart Appliances and Lighting

The evolution of smart appliances and lighting has been a boon for off-grid living. LED lighting, known for its low power consumption, can be further optimized with intelligent controls that adjust brightness based on the time of day or presence in a room. Similarly, energy-efficient appliances can be programmed to operate during periods of high power availability, such as when your solar panels are in full swing. Highlights include:

- Refrigerators that enter a power-saving mode when batteries are low.

- Washing machines that only start when your system's energy production is peaking.

Integrating intelligent appliances into your off-grid home reduces energy consumption. It adds a layer of convenience that modern living demands.

Integration and Control

The true magic happens when these systems—energy, water, and appliances—are smart and integrated. Imagine a home where the energy management system communicates with your water purifiers to run during periods of excess power or where your lighting adjusts not just to motion but to the overall energy status of your home. This level of integration is achieved through:

- Centralized control panels or apps that provide real-time data and remote control over your home's systems.

- Automated routines that, for example, can prioritize energy usage based on your daily schedule or weather conditions.

This interconnectedness ensures that each component of your off-grid home works together, creating an ecosystem that's efficient, responsive, and attuned to your needs.

Wrapping It Up...

As we wrap up this exploration of smart off-grid homes, it's clear that integrating modern technology with traditional off-grid principles opens a new chapter in sustainable living. By embracing automation, we're not just making our lives easier but also reinforcing our commitment to efficiency and environmental stewardship. These advancements are a testament to human ingenuity and our ability to live comfortably, even luxuriously, without leaving a heavy footprint on the planet. As we move forward, let's carry these lessons into the broader context of living in harmony with the world around us, fully aware that every innovation, every system, and every choice is a step towards a more sustainable future.

Keeping The Game Alive

Now that you have equipped yourself with all the tools to thrive in the world of Off-Grid Power, perhaps you can pay it forward and share your newfound knowledge with others. By leaving your honest opinion of this book on Amazon, you provide valuable feedback and guide fellow off-grid enthusiasts towards resources they need for their off-grid power needs.

Thanks for your contribution. Your review plays an important role in spreading awareness to others for their own journey to off-grid power independence.

Simply scan the QR code below to leave your review:

Conclusion

We've been on quite a journey together, exploring various topics, from solar power to wind energy and discovering the essentials of living off the grid. This book has shown us how to install, maintain, and troubleshoot our systems and how to prepare ourselves financially and legally for off-grid living.

Throughout this journey, we've learned some important things that have become our guiding principles. We've understood the value of becoming self-sufficient, caring for the environment, and the joy of seeing our utility bills reduced to nothing. This book has provided practical, step-by-step instructions so that everyone can confidently move towards off-grid independence regardless of their technical skills or budget.

Now, it's your turn to take action. You can start by conducting an energy audit, figuring out where to place your first solar panel, or connecting with other off-gridders. There is no one-size-fits-all approach to off-grid living, and I hope this book can help you make informed decisions that align with your goals.

Remember that the world of renewable energy and off-grid living is constantly evolving. Stay curious, keep learning, and be ready to adapt to new innovations and regulations. Building a system is not just about creating a mindset of adaptability and continuous growth.

I encourage you to share your journey with the off-grid community, whether it's through online forums, local meetups, or conversations with your curious neighbors. Every challenge you overcome, and every success you celebrate adds to our collective knowledge and inspiration.

So, armed with your new knowledge and roadmap, take that first step, even if it's small. The road to off-grid living may be filled with obstacles. Still, the rewards of independence, harmony with nature, and self-sufficiency are priceless. Remember that every small step you take is a giant leap towards a sustainable future for yourself and our planet.

You're not just paving a path to independence; you're lighting the way for others to follow. Let's create a future powered by the sun and the wind and our collective will to make a positive impact.

References

10 benefits of composting toilets for living off-grid. (2023, September, 22). Trelino. https://mytrelino.com/blogs/news/10-benefits-of-composting-toilets-for-living-off-grid

5 solar installation best practices you need to know. (2023, February 3). DuraLabel. https://resources.duralabel.com/articles/5-solar-installation-best-practices-you-need-to-know

9 essential off grid skills every homesteader should know. (2022, January 4). Little Dog Ranch. https://www.littledogranch.com/post/9-essential-off-grid-skills-every-homesteader-should-know

A complete guide to DIY solar panels. (2024). MarketWatch. https://www.marketwatch.com/guides/solar/diy-solar-panels/#:~:text=Based%20on%20our%20research%2C%20you,with%20professional%20installation%20is%20%2417%2C095.

Barnes, S. (2023, December 18). *The top off grid living myths debunked*. The Off Grid Cabin. https://theoffgridcabin.com/the-top-off-grid-living-myths-debunked/

Bidirectional charging and electric vehicles for mobile storage. (n.d.) Energy.gov.

BloombergNEF. (2023, November 22). *Lithium-Ion battery pack prices hit record low of $139/kWh.* BloombergNEF. https://about.bnef.com/blog/lithium-ion-battery-pack-prices-hit-record-low-of-139-kwh/#:~:text=New%20York%2C%20Novem ber%2027%2C%202023,research%20provider%20BloombergNEF%20(BNEF).

Brooks, A. & Saddler, L. (2024, March 7). *What is the best angle and orientation for solar panels?* Forbes https://www.forbes.com/home-improvement/solar/best-angle-for-solar-panels/#:~:text=The%20proper%20solar%20panel%20orientation,the%20most%20amount%20of%20electricity.

Creighton, R. (2012, November 28). *The benefits of airborne wind energy.* IEEE Spectrum. https://spectrum.ieee.org/the-benefits-of-airborne-wind-energy

David, L. (2024a, April 2). *Guide to solar batteries: Are they worth it? (April 2024).* Market Watch. https://www.marketwatch.com/guides/solar/solar-batteries-guide/

David, L. (2024b, September 20). *How much does an off-grid solar system cost? (2024).* Market Watch. https://www.marketwatch.com/guides/solar/off-grid-solar-system-cost/#:~:text=Off%2DGrid%20Solar%20Equipment%20Costs,panels%2C%20inverter%20and%20charge%20controller.

DIY solar panel system installation guide: Step by step. (2023, November 5). Shop Solar. https://shopsolarkits.com/blogs/learning-center/diy-solar-panel-installation-guide

DIY solar water heater. (2023, June 2). Family Handyman. https://www.familyhandyman.com/project/diy-solar-water-heating/

Do-it-yourself home energy assessments. (n.d.). Energy.gov. https://www.energy.gov/energysaver/do-it-yourself-home-energy-assessments

Duval, G. (2024, April 7). *What's the difference between vertical and horizontal wind turbines.* Today's Homeowner.

Energy management and storage for off-grid homes. (2023, February 16). FranklinWH. https://www.franklinwh.com/blog/energy-management-and-storage-for-off-grid-homes

Environmental impact assessment of wind turbines. (n.d.) Collington Winter. micro-https://collingtonwinter.com/blog/environmental-impact-assessment-of-wind-turbines/

Gorr-Pozzi, E., Olmedo-Gonzales, J., & Silva, R. (2022). Deployment of sustainable off-grid marine renewable systems in Mexico. *Frontiers in Energy Research, 10.* rg.2022.1047167

Homeowner's guide to going solar. (n.d.) Energy.gov. https://www.energy.gov/eere/solar/homeowners-guide-going-solar

How do wind turbines work? (n.d.). Energy.gov.

How many solar panels do you need? (n.d.) SunPower.

How much do solar panels cost? | See up to date prices. (n.d.). SolarQuotes. https://www.solarquotes.com.au/panels/cost

How much electricity does an American home use? (n.d.) EIA.

How to size a solar system: Step-by-step. (2020, July 14). Unbound Solar. https://unboundsolar.com/blog/how-to-size-solar-system

Installing and maintaining a small wind electric system. (n.d.). Energy.gov.

Installing and maintaining a small wind electric system. (n.d.). Energy.gov. https://www.energy.gov/energysaver/installing-and-maintaining-small-wind-electric-system#:~:text=Sizing%20Small%20Wind%20Turbines&text=A%201.5%2Dkilowatt%20wind%20turbine,size%20turbine%20you'll%20need.

Is living off the grid now a crime? (n.d.). Off Grid Survival. https://offgridsurvival.com/livingoffthegridcrime/

Jones, K. (2023, October 24). *Internet of things (IOT): A game-changer for energy management.* Medium.

Kindberg, L. (n.d.). *Micro-hydro power: A beginners guide to design and installation.* NCAT ATTRA. https://attra.ncat.org/publication/micro-hydro-power-a-beginners-guide-to-design-and-installation/

Lam, J. & Gobler, E. (2024, January 22). *Solar panel maintenance guide: Keep your panels clean.* CNET. https://www.cnet.com/home/energy-and-utilities/solar-panel-maintenance-guide-how-to-clean-and-repair-solar-panels/

Lozanova, S. (2024, February 23). *How to get a solar permit.* GreenLancer. https://www.greenlancer.com/post/how-to-get-solar-permit#:~:text=Obtaining%20a%20solar%20permit%20may,and%20cost%20might%20be%20sufficient.

Mack, E. (2023, September 7). *Three years in, the biggest benefits and struggles of life off-grid surprise me.* CNET. https://www.cnet.com/home/energy-and-utilities/three-years-in-the-biggest-benefits-and-struggles-of-life-off-grid-surprise-me/

Marsh, J. (2022, May 3). *Geothermal heat pump cost breakdown.* Energy Sage. https://www.energysage.com/heat-pumps/costs-benefits-geothermal-heat-pumps/

Odogwu, C. (2024, January 22). *Monocrystalline vs. polycrystalline solar panels: What's the difference?* CNET. https://www.cnet.com/home/energy-and-utilities/monocrystalline-vs-polycrystalline-which-solar-panels-are-right-for-you/

Perovskite solar cells. (n.d.). Energy.gov. https://www.energy.gov/eere/solar/perovskite-solar-cells

Perry, C. & Tynan, C. (2024, January 10). *How much do off-grid solar systems cost in 2024?* Forbes.

Pickerel, K. (2021, April 19). *Going off-grid in the 2020s: Updated battery choices for today's power needs*. Solar Power World. https://www.solarpowerworldonline.com/2021/04/off-grid-batteries-for-todays-power-needs/

PWM vs MPPT solar charge controllers. (n.d.) Solarcraft. https://www.solarcraft.net/resources/articles/pwm-vs-mppt-solar-charge-controllers

Renewable energy insurance. (n.d.) Liberty Speciality Markets. https://www.libertyspecialtymarkets.com/product/renewable-energy-insurance

Residential clean energy credit. (n.d.). Internal Revenue Service. https://www.irs.gov/credits-deductions/residential-clean-energy-credit

Rinkesh. (2022, July 27). *Various pros and cons of biomass energy*. Conserve Energy Future. https://www.conserve-energy-future.com/pros-and-cons-of-biomass-energy.php

Rosales, P. (2024, February, 13). *Off grid living: Long-term cost comparison*. SunsatStar. https://sunsatstar.com/off-grid-living-long-term-cost-comparison/

Rural energy for America program renewable energy systems & energy efficiency improvement guaranteed loans & grants. (n.d.). Rural Development. https://www.rd.usda.gov/programs-services/energy-programs/rural-energy-america-program-renewable-energy-systems-energy-efficiency-improvement-guaranteed-loans

Scheckel, P. (2018, June 29). *Assess your site for home wind power*. Mother Earth News. https://www.motherearthnews.com/sustainable-living/renewable-energy/assess-site-home-wind-power-zm0z18aszsphe/

Solar Power World. (2022, April 4). *The solar PV system troubleshooting checklist*. Solar Power

World. https://www.solarpowerworldonline.com/2022/04/the-solar-pv-system-troubleshooting-checklist/

Solar PV electrical safety. (n.d.) Electrical Safety Foundation. https://www.esfi.org/solar-pv-electrical-safety/

Standby power: What is it and how can you prevent it? (2023, August 4). Sensorfact - Smart Monitoring for Industry. https://www.sensorfact.eu/blog/standby-power-what-is-it-and-how-can-you-prevent-it/#:~:text=Standby%20consumption%20is%20the%20energy,consume%20power%20at%20that%20moment.

Stringer, A., Vogel, J., Lay, J., & Nash, K. (2017). *Design of rainwater harvesting systems in Oklahoma.* Oklahoma State University. https://extension.okstate.edu/fact-sheets/design-of-rainwater-harvesting-systems-in-oklahoma.html

Takemura, A. F. (2023, July 18). *What's a home energy audit – and should you get one?* Canary Media.

Torres, J. F. and Petrakopoulou, F. (2022). A closer look at the environmental impact of solar and wind energy. *Global Challenges,* 6(8). https://www.ncbi.nlm.nih.gov/pmc/articles/PMC9360340/

Troubleshooting wind turbine problems. (n.d.) Leading Edge Power. https://www.leadingedgepower.com/support/help-with-wind-turbines/troubleshooting.html

Vuković, D. (2024, February 14). *The off-grid laws of every state in America.* Primal Survivor. https://www.primalsurvivor.net/living-off-grid-legal/

Why solid-state batteries are the future of battery technology. (n.d.). Petro Online. https://www.petro-online.com/news/fuel-for-thought/13/international-environmental-technology/why-solid-state-batteries-are-the-future-of-battery-technology/61521

Wikipedia contributors. (2024, March 18). *Small wind turbine.* Wikipedia. https://en.wikipedia.org/wiki/Small_wind_turbine

Wind turbine maintenance: Components, strategies, and tools. (2021, September 28). Enerpac. https://blog.enerpac.com/wind-turbine-maintenance-components-strategies-and-tools/

Zientara, B. (n.d.). *Exciting new solar technologies that actually matter (and why they matter).* Solar Reviews. https://www.solarreviews.com/blog/solar-panel-technologies-that-will-revolutionize-energy-production

www.ingramcontent.com/pod-product-compliance
Lightning Source LLC
Chambersburg PA
CBHW052140070526
44585CB00017B/1911